Annie Wong

THE MAGNIFICENT 9

開門9件事 從基本調味料煮起

basic seasonings for wonderful dishes

ar • salt • wine • peppercorn • fish sauce • oyster sauce
oy sauce • vinegar • sesame oil • sugar • salt • wine •
ppercorn • fish sauce • oyster sauce • soy sauce • vinegar
same oil • sugar • salt • wine • peppercorn • fish sauce
ster sauce • soy sauce • vinegar • sesame oil • sugar
alt • wine • peppercorn • fish sauce • oyster sauce
sauce • vinegar • sesame oil • sugar • salt • wine •

Preface

我從事烹飪導師多年，很幸運這份工作亦
是我的興趣所在，因此腦筋無時無刻都在
思考有關烹飪的事情。

很多朋友及粉絲都以為作為一個熱愛煮食
的人，我家一定擺滿了琳瑯滿目的煮食材
料。我也希望如此。試問一位愛入廚的人
看見心儀的食材，都會巴不得搬回家調製
一番！但奈何香港家居地方淺窄，容不下
太多東西，所以最終在我的廚櫃裏只能存
放一些常用基本調味料。

但你可別看輕這些基本調味料。今次我為
大家介紹的 9 種調味料，只要運用得宜加
上一些心思，便可烹調出數十款美食。

如你也是地方有限，又或你是新手入廚，
正要為你的廚房添置材料，不妨就先由這
9 種調味料開始，保證已足夠令你大快朵
頤！

People ask why I still have such great passion for cooking after being a cookery instructor for so many years. It is because the art of preparing food simply fascinates me. Cooking is a creative art which has no boundaries. It will take you as far as your imagination can go. I have and always will find satisfaction and excitement in preparing food and creating new cuisines.

All my friends and fans know I love to cook. They also think I have a kitchen full of seasonings, enough for all kinds of cuisines. Or like any food enthusiast, always stocking up on the latest trends in cooking material. This is not the case, as most kitchens in Hong Kong are small and have limited storage space. The truth is I stock only the very basic seasonings.

In this cookbook, I will show you how to create wonderful dishes with just 9 basic seasonings. Do not underestimate the power of these seasonings, for they can make a master chef out of you. So if you have the same storage limitations in your kitchen, or you are just a beginner who wants to experience the joy of cooking, why not equip yourself with these 9 basic seasonings. I assure you that you can surprise your family and friends with an array of scrumptious dishes good enough for a feast!

CONTENTS

目錄

QR code

蠔油
Oyster sauce

蠔油是一種很特別的調味料。它的味道並不濃烈，當中蘊含的鮮、鹹、甜味都是淡淡的。雖然如此，加入了蠔油的菜式，總會給人煥然一新的感覺，無論與素菜、肉或海鮮的配搭都得宜。

Oyster sauce has a very unique taste. It has a well-balanced flavour of freshness, saltiness and sweetness. It infuses well into vegetarian, meat and seafood dishes, giving them a deep-reddish brown colour and a fuller flavour.

台式皮蛋肉鬆豆腐
Chilled Tofu with Century Egg and Pork Floss

要選一道簡單、美觀、好味兼百吃不厭的前菜，「台式皮蛋肉鬆豆腐」必然是我的首選！雪白的即食豆腐鋪上皮蛋粒，灑上剁碎芫茜、葱花及肉鬆，單是賣相就攞滿分！其簡易製法更是急救菜之選！

"Chilled Tofu with Century Egg and Pork Floss" is my choice for a simple, yet irresistible starter. The contrasting colours of the white tofu and the black century egg, along with toppings of chopped coriander, spring onion and pork floss, make the presentation very elegant! Furthermore, this dish can easily be prepared in minutes.

材料

嫩豆腐	1 盒（雪凍）
皮蛋	1 個（切碎）
即食肉鬆	2-3 湯匙
蔥碎	1 湯匙
芫茜碎	1 湯匙（隨意）

調味醬油

蠔油	3 湯匙
靚麻油	2 湯匙
糖	1 茶匙

做法

1. 調味醬油拌勻，試味。

2. 豆腐濾去水分，盛碟內，用刀別 8 份。

3. 把皮蛋放豆腐上，淋上調味醬油，灑適量肉鬆、蔥及芫茜碎在面成涼拌前菜。

Ingredients

1 packet soft tofu, chilled in fridge
1 century egg, roughly chopped
2-3 tbsp pork floss
1 tbsp chopped spring onion
1 tbsp chopped coriander, optional

Seasoning sauce

3 tbsp oyster sauce
2 tbsp sesame oil
1 tsp sugar

Method

1. Mix seasoning sauce together, adjust taste.

2. Drain tofu, arrange on a plate and cut into 8 portions.

3. Arrange century egg on top of the tofu, pour seasoning sauce over, and sprinkle with pork floss, spring onion and coriander. Serve as a starter.

Note

· 調味醬油味要重，因要配合味道偏淡的豆腐。
· 如採用盒裝豆腐，應在食用前才從雪櫃取出，可避免豆腐瀉水。

· The seasoning sauce is designed to be rich and flavourful, so it can enhance the bland taste of the tofu.
· Chill the tofu in the original carton in the fridge. Take out the tofu just before serving. This will prevent the loss of moisture from the tofu.

蠔油牛肉
Fried Beef Fillet with Oyster Sauce

蠔油牛肉，這道家傳戶曉的小菜，向來都是大廚師的考牌菜。因為成功的炒牛肉，必需要有鑊氣。牛肉當然要鮮嫩適中，還要不多不少的芡汁以及仍保持青翠的菜蔬。要達如此要求，真是少點功夫都不成！

"Fried Beef Fillet with Oyster Sauce" is a dish that is easy to cook but difficult to master. The beef should be tender with a delicate charred aroma. There should be just enough sauce to linger on the beef, no more, no less. The vegetables should be crispy and jade green in colour. To cook it to perfection, even the master chef will be put to the test!

材料

嫩牛肉	150 克
芥蘭	300 克
薑	2 片
蒜頭	1 粒（切片）

醃料

（一）

梳打食粉	1/4 茶匙
水	1 湯匙

（二）

生抽	1 1/2 茶匙
糖	1/4 茶匙
麻油及胡椒粉	各少許
生粉	1/2 茶匙
水及油	1 湯匙

炒菜調味料

鹽	1/3 茶匙
糖	1/2 茶匙
水	3 湯匙

芡汁

水	5 湯匙
蠔油	2 湯匙
生抽	1 茶匙
糖	1/2 茶匙
生粉	1 茶匙

Ingredients

150g beef fillet or beef flank
300g Chinese kale
2 slices ginger
1 clove garlic, sliced

Marinade

(A)

1/4 tsp cooking soda
1 tbsp water

(B)

1 1/2 tsp light soy sauce
1/4 tsp sugar
a little sesame oil and pepper
1/2 tsp potato starch
1 tbsp each water and oil

Seasonings for vegetables

1/3 tsp salt
1/2 tsp sugar
3 tbsp water

Sauce

5 tbsp water
2 tbsp oyster sauce
1 tsp light soy sauce
1/2 tsp sugar
1 tsp potato starch

Note

· 油鑊內先加炒菜調味料，然後才加菜一起炒，這樣可避免菜於淨熱油內炒乾或燶，又可使菜均勻吸取調味。

· Add seasonings for vegetables first before the vegetables. Toss them together. This will prevent vegetables from being charred in the hot oil, and will ensure the seasonings are well absorbed.

做法

1. 牛肉橫紋切薄片，先與醃料（一）拌勻，待梳打食粉水滲透肉裏，再加醃料（二）拌勻，待 30 分鐘。

2. 芥蘭摘去白色花，切段，沖淨及隔乾。

3. 燒 2 湯匙油爆香薑片，加炒菜調味料拌勻，隨即放入菜段，用中火兜炒，可蓋鑊蓋 30 秒至 1 分鐘至菜熟。取起菜，瀝淨便可上碟。

4. 鑊洗淨抹乾，加 2 湯匙油，放下牛肉，弄散，每邊烘一烘，加入蒜片一起兜炒至牛肉九成熟。可把牛肉取出或撥至鑊邊。

5. 倒入芡汁煮滾，牛肉放回芡汁內，快手兜炒，把牛肉及芡汁淋於芥蘭上。

Method:

1. Slice beef thinly across the grains, mix with marinade (A) until well absorbed, add (B) and mix well. Set aside for 30 minutes.

2. Trim Chinese kale, cut into sections, rinse and drain.

3. Heat 2 tbsp oil in the wok, sauté ginger until fragrant, add seasonings for vegetable, stir well, add Chinese kale, toss over medium heat, cover and cook for 1/2 to 1 minute until vegetables are just cooked. Remove and drain well. Arrange on a plate.

4. Start with a clean wok, heat 2 tbsp oil, lower sliced beef into hot oil, loosen with chopsticks, fry each side until light golden brown. Add garlic and fry together until beef is almost done. Remove beef or push to the sides of the wok.

5. Add sauce to wok and bring to a boil. Return beef to sauce and fry together. Transfer beef with sauce on top of the Chinese kale and serve.

蠔皇原隻鮑魚
Braised Abalone in Oyster Sauce

近年乾鮑魚價格不停上漲，除非付天價，否則都不能在酒樓食到有水準的原隻蠔皇乾鮑。與其白花了錢，不如乾脆自家做，過程並非想像中的繁複，而且多出來的鮑汁更可烹調其他美味菜式，一舉數得！

The price of dried abalone has risen a great deal in recent years. It costs much more to savour good ones in restaurants nowadays. At these prices, why not cook them at home? The preparation and cooking are really not as tedious as you think. The extra braised abalone sauce can be used to prepare other delicious dishes too!

材料

12 隻	乾鮑魚（* 日本吉品鮑魚 30 至 33 頭）

煨鮑魚料

（一）

薑	10 片
葱	4 條
水	蓋面

（二）

光雞	1 隻（斬成 8-10 件）
豬扒	4 件（600 克）（枚頭豬扒連骨）
水	蓋面

扣鮑魚汁料

乾葱	2 粒（略拍）
煨鮑魚濃湯	2 杯
水	1 杯
蠔油	4 湯匙
生抽	2 湯匙
糖	2 湯匙

馬蹄粉水

馬蹄粉	適量
水（拌馬蹄粉）	適量
老抽	適量（調色）

做法

1. 乾鮑魚的處理方法：
   乾鮑魚沖淨，用蓋面水浸 4 小時至略軟身。用牙刷輕輕洗擦乾鮑魚表面，沖淨後放瓦煲內，注入水蓋過面，加薑及蔥；先煮滾，改用中慢火，蓋好，煮 1 1/2 小時。熄火後不可打開煲蓋焗至水凍。取出鮑魚，用小刀小心清理一端的內臟，再沖淨。

2. 光雞及豬扒出水，盡量避免有血水，沖淨備用。

3. 煨鮑魚：
   瓦煲內放一塊竹撻，先放一層豬扒墊底，排上鮑魚，把雞件鋪面，倒入熱水至剛剛蓋面，再煮滾，改用慢火（只見小泡泡於水面），蓋好，熬 2-3 小時，關火。原煲蓋好，待至煨鮑魚濃湯涼透。（不需放雪櫃）
   重複慢火煨鮑魚做法，每次開火前先注入熱水蓋過鮑魚，並查看有沒有黏底。
   重複 3-4 次至鮑魚腍身（視乎鮑魚大小）。取出鮑魚，留起煨鮑魚的濃湯。

4. 扣鮑魚方法：
   燒熱 1 湯匙油，爆香乾蔥，倒入汁料煮滾，熬 10 分鐘。放下鮑魚，改用中慢火扣 2 小時，原煲蓋好至汁涼透。

5. 享用時，鮑魚與汁煮滾，拌入適量馬蹄粉水埋芡，用適量老抽調色，趁熱品嘗。

Remarks

選購乾鮑魚是以「頭」計算。例如 30 至 33 頭，這是代表鮑魚的大細，每 30 至 33 隻鮑魚共 600 克重，約 18-20 克一隻。鮑魚越大，頭數越少。

Abalones are categorized by size. Abalones of similar size are sold in 600g lots. When there are 30 to 33 abalones in a lot, depending on their size, each abalone weights between 18 to 20g. The bigger the abalone, the fewer you will get in a lot.

Ingredients

12 dried abalones
(*30 to 33 Japanese Kippin abalones
per 600g)

Braising sauce

(A)
10 slices ginger
4 stalks spring onion
water to cover the abalones

(B)
1 chilled chicken, cut into 8-10 pieces
4 thick slices pork chop with bone (600g)
water to cover the abalones

Sauce

2 shallots, lightly crushed
2 cups braising stock
1 cup water
4 tbsp oyster sauce
2 tbsp light soy sauce
2 tbsp sugar

Thickening

water chestnut starch
water to blend
a little dark soy sauce for colour

Method

1. **Curing dried abalone:**
   Rinse abalones, soak in water for 4 hours until a little softened. Lightly clean the surface of the abalone with a tooth brush. Rinse again and place in a pot. Add water to cover, add ginger and spring onion. Bring to a boil, lower heat, cover and simmer for 1 1/2 hours . Do not open lid until water has completely cooled down. Remove abalones, using a small sharp knife, remove the entrails from one end of the abalone carefully. Rinse again.

2. Blanch chicken pieces and pork chop well to avoid any blood draining out. Rinse well before use.

3. **Braising abalone:**
   Use a clay pot, line the inside of the pot with a bamboo mat. Lay alternate layers of pork chop, abalones and chicken pieces, add hot water to cover, bring to a boil, reduce heat to simmering (only very small bubbles should appear on the water surface.), cover with lid and simmer for 2 to 3 hours.
   Switch off heat. Do not open the pot, set aside until abalone stock has cooled down. (Do not put in the fridge)
   Repeat the simmering process 3 to 4 times until abalones are tender. Each time before switching on the heat, check to make sure no meat is sticking to the bottom; add enough water to cover the abalones. When abalones are tender, remove and save the stock.

4. **Seasoning abalones:**
   Heat 1 tbsp oil, sauté shallots until fragrant, add sauce ingredients and bring to a boil. Simmer for 10 minutes. Add braised abalones into sauce and simmer for 2 hours until well flavoured. Switch off heat. Do not open the pot. Set aside until cool.

5. To serve: Bring abalones and sauce to a boil. Stir in enough thickening to form gravy. Serve hot.

Note

· 把鮑魚夾在豬扒與雞中央,除可避免鮑魚黏底焦燶,也可避免鮑魚浮面而變乾。

· 每次開火燜鮑魚前,要用筷子轉竹笪一圈,可避免燜時材料黏底。

· 我燜乾鮑魚會用斬成塊、帶有骨及肥肉的枚頭豬扒,方便鋪平為鮑魚墊底避免焦燶。豬扒的骨及肥肉,味道較淨肉香濃,有助提升鮑魚鮮味。

· 雞可用冰鮮,較新鮮雞經濟,必須連骨及皮。還可買多 8 至 10 隻雞腳一同燜,可提供骨膠原,使燜後的鮑魚更香滑。

· Abalones are sandwiched between pork chop and chicken pieces to prevent them from sticking to the bottom of the pot, or surfacing above the stock and becoming dry.

· Every time before switching on the heat, use chopsticks to swirl the bamboo mat round several times to check if there is any meat sticking on the pot.

· I choose semi-fat pork chop with bones and fat, chopped into pieces for putting underneath the abalones to prevent them sticking and burning. Both bones and fat enrich the taste of the abalones.

· Use chilled chicken which is more economical than fresh chicken. Braise together with its bones and skin. You can even add 8 to 10 more chicken feet to the pot. Their gelatine enhances the flavour and texture of the braised abalones.

紅燒菇片
Braised Mushroom Slices

物盡其用是我一貫的宗旨，有剩餘的鮑汁，我又怎會不好好利用呢？其中一個嘗試就是用鮑汁煨雞髀菇，效果一流，甚至充當宴席上的「蠔汁素鮑脯」亦毫不失禮！

My principle for food is "To get the most out of it." So I use the extra braised abalone sauce to braise shaggy ink capped mushrooms. Surprisingly, the mushrooms taste just like abalones. I highly recommend this dish for the vegetarian menu.

材料

雞髀菇	500 克
西生菜	1 個
蒜頭	2 粒（略拍）
乾葱	1 粒（略拍）

汁料

水	1 杯
蠔油	2 湯匙
生抽	1 茶匙
糖	1 茶匙
麻油及胡椒粉	各少許

生粉水

生粉	1 茶匙
水	1 湯匙

做法

1. 西生菜撕開一片片，沖淨。放半鑊水內，加 1 茶匙鹽、糖及 1 湯匙油煮滾，灼熟西生菜，取出隔淨，放在碟上。

2. 雞髀菇沖淨及切去枯黃部分，切 0.5 厘米厚片，形似鮑脯。

3. 燒熱 3 湯匙油，放下蒜頭及菇片炒透，取出．

4. 再燒熱 1 湯匙油，爆香乾葱，倒入汁料煮滾。

5. 菇片放回汁料燜煮 5-8 分鐘至入味及上色，最後用適量生粉水埋芡，盛在西生菜上。

Ingredients

500g shaggy ink cap mushrooms
1 lettuce
2 cloves garlic, lightly crushed
1 shallot, lightly crushed

Sauce

1 cup water
2 tbsp oyster sauce
1 tsp light soy sauce
1 tsp sugar
a little sesame oil and pepper

Thickening

1 tsp potato starch
1 tbsp water

Method

1. Cut lettuce into pieces, rinse well. Bring 1/2 wok of water to a boil, add 1 tsp salt, 1 tsp sugar and 1 tbsp oil. Cook lettuce, remove and drain well. Arrange on a plate.

2. Rinse and trim mushrooms. Cut slantingly into 0.5cm thick slices, resembling sliced abalone.

3. Heat 3 tbsp oil, sauté garlic and mushroom slices until cooked. Remove.

4. Heat 1 tbsp oil, sauté shallot until fragrant, add sauce and bring to a boil.

5. Return mushroom slices to sauce, braise for 5 to 8 minutes until well flavoured and coloured. Lastly thicken sauce with enough potato starch mixture. Pour the mushroom slices onto the lettuce.

Note

· 宜選購較粗身的雞髀菇,斜斜橫切0.5厘米厚片成鮑脯形狀。雞髀菇肉質爽脆,經燜煮後只會更加入味而不會霉爛。

· 以紅燒方法烹煮成素鮑脯,味道與口感可媲美鮑脯。

· Choose thick cylindrical shaped shaggy ink cap mushrooms. Cut slantingly into 0.5cm thick slices to resemble sliced abalone. Shaggy ink cap mushrooms have a crunchy texture. They will become more flavoursome after braising and will not turn mushy easily.

· This is a very nice substitute for braised abalone in a vegetarian menu.

魚露
Fish Sauce

使用魚露作調味料，主要流行於中國南部沿岸以及東南亞一帶，例如潮州菜便經常見它的踪影。

魚露味道鹹香味鮮，特別適合海鮮烹調。用它入饌，能帶來一絲異鄉味道，為菜式加添新鮮感。

現時有些魚露在包裝上會寫上一個數字度數，到底代表甚麼呢？那其實是表示魚露內魚汁的比例，度數愈大，魚鮮味愈濃。

Fish sauce is a dominant seasoning along the coastal areas of southern China and Southeast Asian.

It plays a prominent role in the Chaozhou cuisines. The fresh-saltiness of anchovy blends very well with seafood. Dishes with fish sauce always give you a touch of exoticism.

What does the number on the bottle of the fish sauce mean?

It states the percentage of fish essence in the fish sauce. The higher the number, the richer is the flavour.

魚露酸辣汁
Spicy Fish Sauce Vinaigrette

材料		Ingredients
魚露	3 湯匙	3 tbsp fish sauce
青檸汁	2 湯匙	2 tbsp Thai lime juice
白醋	1 湯匙	1 tbsp distilled white vinegar
糖或椰糖	2 茶匙	2 tsp sugar or palm sugar
指天椒	1-2 隻（切碎）	1-2 bird's eye chillies, chopped
蒜茸	2 茶匙	2 tsp chopped garlic
葱粒	1 湯匙	1 tbsp chopped spring onion
芫茜碎	1 湯匙	1 tbsp chopped coriander
凍開水	適量（隨意）	cooled boiled water, optional

做法

魚露酸辣汁拌勻及試味。

Method

Blend spicy fish sauce vinaigrette, adjust taste to your liking.

魚露乳鴿
Pigeon in Fish Sauce

材料

冰鮮乳鴿	1 隻
薑	2 片（略拍）
乾葱	2 粒（略拍）
魚露	1/3-1/2 杯
黃酒	3 湯匙
冰糖	3 湯匙
水	5 杯

做法

1. 乳鴿沖淨，放入半鍋滾水內汆水 3-5 分鐘，取出沖淨，抹乾。

2. 用一個中型鍋，燒 3-4 湯匙油，爆香薑及乾葱，放下乳鴿，用中火熱油煎至鴿皮呈微黃色。加水至蓋過乳鴿，再加魚露、黃酒及冰糖。煮滾後，改用中慢火，蓋上鍋蓋燜熟，每 10 分鐘用湯杓取汁淋勻乳鴿，使均勻上色。

3. 乳鴿約 30 分鐘便可煮熟。熄火後，留乳鴿於汁內多 10 分鐘才取出。斬件上碟，淋少許魚露汁在面。

Ingredients

1 chilled pigeon
2 slices ginger, lightly crushed
2 shallots, lightly crushed
1/3-1/2 cup fish sauce
3 tbsp brown wine
3 tbsp rock sugar
5 cups water

Method

1. Clean and rinse pigeon, blanch in half pot of water for 3-5 minutes, remove and rinse again. Wipe dry.

2. In a medium size pot, heat 3 to 4 tbsp oil, sauté ginger and shallots, lower pigeon into medium hot oil and fry until skin appears golden. Add enough water to cover, add fish sauce, brown wine and rock sugar. Bring to a boil, reduce heat to medium-low, cover and simmer. Baste pigeon with the sauce every 10 minutes to enhance the colouring.

3. When pigeon is cooked after 30 minutes, switch off heat, leave pigeon in the sauce for another 10 minutes. Take out and chop into pieces. Pour a little sauce over and serve.

Note

· 不同牌子的魚露鹹味與色度都不同，須試味來調校魚露與冰糖的份量。

· 將乳鴿汆水，可去腥和去血水，讓食味更佳和汁料顏色清澈。

· The saltiness and colour of fish sauce differ from brand to brand. Adjust the quantity to be used accordingly and balance the taste with rock sugar.

· To prevent scum formation, blanch the pigeon in boiling water to remove the excess blood.

烤豬頸肉沙律
Grilled Pork Cheek Salad

材料

豬頸肉	1 件（250 克）
魚露	2-3 湯匙
青檸汁	1 湯匙
糖	1 茶匙
魚露酸辣汁（做法看第 25 頁）	

配菜

洋蔥（白洋蔥 或 紅洋蔥）	1/2 個（切條）
生菜葉	2-3 塊（撕成小塊）
指天椒	1-2 隻（切碎）
紅辣椒碎	1 湯匙
乾蔥	2 個（切片）

撒面材料

焗香花生碎	2 湯匙

做法

1. 把魚露、青檸汁及糖搽勻豬頸肉醃 1 小時。醃好的豬頸肉放中火烤爐內烤熟，不時把肉件反轉，烤至兩面甘香。取出，待 1-2 分鐘才切片。

2. 配菜切好。

3. 豬頸肉片及配菜齊放大碗內與魚露酸辣汁拌勻。

4. 上碟後多撒些香口的焗香花生碎在面。

Ingredients

1 piece pork cheek (250g)
2-3 tbsp fish sauce
1 tbsp Thai lime juice
1 tsp sugar
spicy fish sauce vinaigrette (refer to P.25)

Side vegetables

1/2 onion (brown or red onion), cut into strips
2-3 lettuce leaves, torn into bite-size pieces
1-2 bird's eye chillies, chopped
1 tbsp chopped red chillies
2 shallots, sliced

Toppings

2 tbsp chopped roasted peanuts

Method

1. Rub fish sauce, Thai lime juice and sugar over pork cheek, set aside for 1 hour. Grill pork cheek in medium-hot grill, turning from side to side until cooked and evenly charred. Remove, rest for 1 to 2 minutes before slicing.

2. Cut all vegetables.

3. Place pork cheek and vegetables in a mixing bowl, mix well with spicy fish sauce vinaigrette.

4. Remove to a platter and sprinkle with chopped roasted peanuts. Serve as starter.

斜刀片切豬頸肉，既可切斷肉纖維，樣子也好看些。

Slice the grilled pork cheek at an angle. This will not only cut through the tissues, but will also make the slices much more presentable.

香煎蠔餅
Baby Oyster Omelette
– Chaozhou Style

材料

蠔仔	150 克
大蛋	4 個
芫茜碎	1 湯匙
葱粒	2 湯匙

蠔仔調味料

魚露	1 茶匙
胡椒粉	少許

蛋調味料

魚露	1 茶匙
胡椒粉	少許

粉漿

番薯粉	3 湯匙
水	6 湯匙
魚露	1/2 茶匙

做法

1. 蠔仔撒少許生粉，用手輕輕洗揉，小心挑出蠔仔殼碎，沖水多次，隔淨及抹乾；加調味料拌勻。

2. 番薯粉、水及魚露拌勻成粉漿。

3. 蛋加調味料拂勻。

4. 燒 3 湯匙油，放入蠔仔炒數下，隨即加粉漿，煎一煎，倒入蛋液，撒下芫茜及葱，輕輕兜炒，以中猛火烘香。反轉另一面，沿鑊邊注入少許油，烘至兩面甘香。

5. 上碟後以胡椒粉及魚露蘸吃。

Ingredients

150g baby oysters
4 large eggs
1 tbsp roughly chopped coriander
2 tbsp diced spring onion

Seasonings for oysters

1 tsp fish sauce
a few shakes of pepper

Seasonings for eggs

1 tsp fish sauce
a few shakes of pepper

Batter

3 tbsp sweet potato starch
6 tbsp water
1/2 tsp fish sauce

Method

1. Lightly rub baby oysters with a little potato starch, discard any broken shell, rinse several times, drain and wipe dry. Mix with seasonings.

2. Blend sweet potato starch with water and fish sauce to a batter.

3. Beat eggs with seasonings.

4. Heat 3 tbsp oil in the wok, fry baby oysters lightly, add batter and fry until almost set. Add beaten eggs, coriander and spring onion, mix gently and fry until golden brown. Turn omelette over, drizzle a little oil round the sides of the omelette, continue to cook until both sides are golden brown.

5. Transfer to a plate, sprinkle a little white pepper on the omelette and serve with fish sauce as a dip.

Note

· 採用番薯粉煎蠔餅,吃時煙韌有質感;次選可用生粉,但勿用粟粉,因黏性不同。

· Sweet potato starch, which has a springy texture, is traditionally used for this batter. If unavailable, potato starch can be a substitute, but not cornstarch.

碌酼豬肉
Authentic Pork Stew with Fish Sauce

材料

五花腩肉連皮	600 克
片糖	1/2 塊（剁碎）
香茅	1 條（略拍）
薑	4 片
魚露	4-5 湯匙
水	

做法

1. 五花腩連皮切 4 厘米闊長條，再橫切成 3 厘米小件。

2. 鍋內燒 2-3 湯匙油，加片糖碎用慢火煮溶，先把腩肉皮放糖漿內慢火煎至金黃色。

3. 加入香茅、薑及魚露拌勻，再加入適量水蓋過肉面。煮滾，改用慢火，蓋好，燜至肉腍；每 15 至 20 分鐘拌一拌肉避免黏鍋底，需要時加入適量水。

4. 肉燜腍後改用中猛火略把汁煮至濃稠。

Ingredients

600g belly pork with skin
1/2 slab brown sugar, chopped
1 stalk lemongrass, lightly crushed
4 slices ginger
4-5 tbsp fish sauce
water

Method

1. Cut belly pork with skin into 4cm long strip. Then cut across into 3cm pieces.

2. Heat 2 to 3 tbsp oil in a pot, add chopped sugar, stir to dissolve over low heat. Add pork pieces with skin side down into the syrup, cook slowly until golden brown.

3. Add lemongrass, ginger and fish sauce, mix well and add enough water to cover the pork pieces. Bring to a boil, reduce heat, cover and simmer over low heat. Stir pork pieces every 15 to 20 minutes to prevent sticking. Add more water if necessary.

4. When pork is tender, bring heat up and fry until sauce becomes gravy.

糖要煮至起泡，才放入豬腩肉慢火煎至皮呈
金黃色。小心避免熱油及糖漿飛濺於手上。

Boil sugar until lightly syrupy before adding
the pork pieces. Fry in low heat until skin is
golden brown. Be careful to avoid the hot oil
and syrup from scalding the hands.

椒鹽九肚魚
Crispy Spiced Bombay Duck

材料

九肚魚	600 克
魚露	2 1/2 - 3 茶匙
胡椒粉	少許
紅辣椒碎	1 茶匙
蒜茸	1 茶匙
淮鹽	1 茶匙
生粉	1 杯

做法

1. 九肚魚沖淨，剪去頭、鰓及肚部位，再沖淨及抹乾；拌入魚露及胡椒粉醃 5 分鐘。

2. 燒 1/3 鑊油，油熱時把九肚魚均勻沾上生粉，逐條放入油內，用猛火熱油炸至熟，取出濾油。

3. 再把鑊內的油燒熱，放下九肚魚多炸一次至金黃香脆，取出隔淨油分。

4. 鑊內燒 1/2 湯匙油，爆香紅辣椒及蒜茸，加入九肚魚，撒下適量淮鹽，一起兜炒均勻，即可上碟。

** 淮鹽做法參看第 60 頁

Ingredients

600g Bombay duck
2 1/2-3 tsp fish sauce
a little pepper
1 tsp chopped red chilli
1 tsp chopped garlic
1 tsp spiced salt
1 cup potato starch

Method

1. Clean and rinse fish, cut off head, gills and stomach area, rinse well and wipe dry. Mix with fish sauce and pepper. Set aside for 5 minutes.

2. Heat 1/3 wok of oil until hot, coat each fish evenly with potato starch, lower them one by one into the hot oil and deep-fry until cooked. Remove and drain.

3. Reheat oil in the wok, deep-fry the fish a second time until golden brown and crispy. Remove and drain off excess oil.

4. Heat 1/2 tbsp oil in the wok, sauté red chilli and garlic until fragrant. Return fried fish to the wok and toss together with enough spiced salt to taste. Serve at once.

**Spiced salt refer to P.60

Note

· 街市賣的九肚魚都是冰鮮的，宜挑選魚身較堅挺、魚眼及魚鰓有光澤的，這表示魚較新鮮。
· 用魚露醃九肚魚，除可讓九肚魚更入味外，魚露的鹹味還可令魚身更挺。
· 九肚魚必須在炸時才沾上生粉，保持乾爽。
· 炸熟的魚再多炸一次，可揮發出更多水分，使魚身較長時間保持酥脆。

· Choose chilled Bombay duck with firm meat, bright eyes and gills.
· The flavour of the fish sauce not only enhances the Bombay duck, it's saltiness also firms up the meat.
· Coating the fish just before deep-frying keep it light and dry.
· Deep-frying the fish a second time will remove the excess moisture and maintain its crispiness longer.

番茄湯蛋餃
Mini Omelettes with Tomato Consommé

Ingredients

3 eggs
150g minced pork
2 tomatoes
2 tbsp spicy Sichuan vegetable julienne
3 cups water
a little coriander and spring onion

Seasonings for eggs

1 tsp fish sauce
a few shakes of pepper

Seasonings for pork

1 1/2 tsp fish sauce
1/4 tsp sugar
1 tsp potato starch
1/2 tbsp chopped coriander
1 tbsp chopped spring onion

Tool

1 metal ladle

材料

蛋	3 個
免治豬肉	150 克
番茄	2 個
榨菜絲	2 湯匙
水	3 杯
芫茜及葱	少許

蛋調味料

魚露	1 茶匙
胡椒粉	少許

肉調味料

魚露	1 1/2 茶匙
糖	1/4 茶匙
生粉	1 茶匙
芫茜碎	1/2 湯匙
葱碎	1 湯匙

用具

金屬湯杓	1 個

Note

· 用湯杓代替鑊煎蛋餃，可讓蛋餃大細均
  一。每次煎蛋前必須把湯杓內則搽勻油，
  否則蛋很容易黏着。

· Using a ladle instead of a wok to fry
  the egg will standardize the size of the
  omelettes. Be sure to glaze the inside
  of the ladle with oil before frying each
  omelette to prevent sticking.

做法

1. 蛋加調味料拌勻，番茄切角，免治豬肉加調味料拌勻成餡料。

2. 取一湯杓，用適量油抹勻內則；慢火燒熱湯杓，倒下 2 湯匙蛋液，
   慢慢盪勻湯杓內則至成圓形蛋皮。蛋液未熟透時加入 1 茶匙肉餡
   於蛋皮之一邊，用另一半蛋皮覆蓋肉餡成半圓形蛋餃，慢火煎至
   金黃；重複做法煎下一個蛋餃。

3. 水放鍋內煮滾，加榨菜絲及番茄一起煮至湯濃香。把煎好的蛋餃
   放湯內煮熟，撒些芫茜及蔥上面成湯菜。

Method

1. Beat eggs with seasonings. Cut tomatoes into chunks. Mix minced pork with seasonings as filling.

2. Grease the inside of the ladle, heat up gently, add 2 tbsp beaten egg, swirl evenly round the sides of the ladle until almost set. Put 1 tsp filling on one side, fold the other half over the filling to form a mini omelette. Fry until golden brown. Repeat this process to make more omelettes.

3. Bring water to a boil, add Sichuan vegetable and tomatoes, simmer until well flavoured. Add mini omelettes to cook. Lastly, sprinkle with coriander and spring onion to serve.

胡椒
Peppercorn

胡椒可說是眾多香料中的王者，它的足跡遍佈中西各地的菜餚，自古便已於香料貿易中佔一重要地位。胡椒散發的香味非常吸引，總能給人一種溫暖窩心的感覺！

白胡椒除可用於烹飪外，用棉紗布袋紮好，放入櫃內可驅蟲。

Peppercorn is the most widely used among all spices. It complements both Chinese and Western cuisines and has been an important commodity since ancient times. The peppery heat wakes up the taste buds, and accents the dish to a lingering finish!

Besides the cooking purpose, put some white peppercorns in a muslin bag, tie the opening with a string and place inside your cupboard as pesticide.

黑白胡椒鹹菜排骨湯
Black and White Peppercorns with Pork Rib Broth

提起潮州菜系的湯羹，立即浮於腦海的便是「胡椒鹹菜豬肚湯」。
要做一個像樣的「胡椒鹹菜豬肚湯」，首要條件便是要有濃重的胡
椒味，愈重愈好，最好是喝後要汗流浹背才過癮！在冬天時飲用此
湯亦有祛寒暖胃的效用，血氣欠佳的人士會登時有暖笠笠的感覺。
若然覺得清洗豬肚繁複，大可用排骨代替，肉味會更香濃。

Among the many Chaozhou broths, 'Peppercorn with Pig's Tripe
Broth' is one of my favourites. The heat from the peppercorn should
be so intense that a sip of the broth will make you sweat. It is a
wonderful comfort broth in the winter for it warms up your body.
I recommend using pork rib instead of pig's tripe, because it is much
less tedious to prepare. The broth will be rich and meaty too!

Ingredients

400g pork ribs, chopped into 4cm sections
120g Chaozhou preserved cabbage
1/2 tbsp black peppercorns, lightly crushed
1/2 tbsp white peppercorns, lightly crushed

Method

1. Rinse pork ribs, blanch in boiling water for 3 to 5 minutes, remove and rinse again.

2. Place black and white peppercorns and 10 cups water in a pot, bring to a boil. Add pork ribs and boil for 5 minutes, reduce heat, cover and simmer over medium-low heat for 30 minutes until it becomes a rich pork broth.

3. Rinse preserved cabbage, cut into bite-size pieces, add to broth and continue to simmer for 15 minutes until well flavoured. Now there should be about 6 cups of broth. Adjust taste. The broth is full of peppercorn aroma. It tastes a bit spicy hot and it is very good for health.

材料

排骨	400 克（斬 4 厘米）
潮州鹹菜	120 克
黑胡椒粒	1/2 湯匙（略壓）
白胡椒粒	1/2 湯匙（略壓）

做法

1. 排骨沖淨，汆水 3 至 5 分鐘，取出沖淨。

2. 鍋內放黑白胡椒粒及 10 杯水煮滾，加排骨，先滾 5 分鐘，改中慢火，蓋好煲 30 分鐘至排骨出味。

3. 鹹菜沖淨，切細件。加入湯內繼續中慢火煲 15 分鐘，剩約 6 碗湯。試味，趁熱享用，這湯可暖胃及祛寒。

Note

· 用白鑊將黑、白胡椒粒先炒香，取出略壓至外殼微微爆烈才用來煲湯，可助辛香味短時間溢出。

· Dry-fry the peppercorns in a wok until fragrant. Remove and lightly crush them so that the spicy hot flavour can be infused into the broth easily.

黑椒牛肉粒
Sauté Beef Cubes
with Black Peppercorn

如果牛肉用作快炒，我喜歡選用肉眼部位，除了肉香外，還因為
這部分的油脂分佈均勻，因此份外軟腍。要做到外香內嫩，最緊
要是先把牛肉粒的外層煎香，然後才加入其他材料快速兜炒。加
上惹味的黑椒，這道菜最適宜配紅酒。

Rib-eye steak is my favourite choice for stir-fry. Its marbled fat
contributes to its tenderness and beefy flavour. Sear the meat first
before adding vegetables, then toss quickly over high heat. The
addition of black peppercorn will further enhance its beefy flavour.
This dish pairs well with red wine.

材料

牛肉眼扒	300 克
洋葱	1/2 個
芥蘭莖粒	1 杯
蒜茸	2 茶匙
黑胡椒碎	1/2 茶匙

醃料

黑胡椒碎	1/4 茶匙
生抽	1 湯匙
糖	1/2 茶匙
生粉	1 茶匙
油	1 湯匙

芡汁

水	5 湯匙
生抽	1 1/2 茶匙
糖	1/2 茶匙
生粉	1 茶匙

做法：

1. 牛肉眼扒切粗粒，與醃料拌勻待 10 分鐘。

2. 洋葱切粗粒。

3. 芥蘭莖切粗粒放滾水內，加少許鹽及糖灼
   1/2 至 1 分鐘，保持脆口，取出隔淨。

4. 燒 2 湯匙油，把牛肉粒炒至八成熟，取出。

5. 剩餘油爆香黑胡椒碎及蒜茸，牛肉回鑊，
   加洋葱兜炒，最後加芥蘭莖粒及芡汁一起
   兜勻。

Ingredients

300g rib-eye steak
1/2 onion
1 cup diced Chinese kale stalk
2 tsp chopped garlic
1/2 tsp crushed black peppercorn

Marinade

1/4 tsp crushed black peppercorn
1 tbsp light soy sauce
1/2 tsp sugar
1 tsp potato starch
1 tbsp oil

Sauce

5 tbsp water
1 1/2 tsp light soy sauce
1/2 tsp sugar
1 tsp potato starch

Method

1. Cut rib-eye steak into large cubes, mix with marinade and set aside for 10 minutes.

2. Cut onion into large cubes.

3. Cut Chinese kale stalks into cubes, blanch in boiling water together with a little salt and sugar for 1/2 to 1 minute, keeping them crispy. Remove and drain.

4. Heat 2 tbsp oil, fry beef cubes until 80% cooked, remove.

5. Sauté crushed black peppercorn and garlic until fragrant, return beef cubes to wok, add onion and fry together. Lastly add Chinese kale and mix well with sauce.

Note

· 胡椒粉的辛香味不及磨碎的胡椒粒。若要爆香配肉及海鮮，或作醬汁，必須選用磨碎胡椒粒。

· Crushed peppercorn offers a stronger and more pungent flavour than ground pepper. Choose crushed peppercorn for frying with meat and seafood.

星洲黑胡椒蟹
Black Pepper Crab Singapore Style

有一次到新加坡旅遊嘗過「黑胡椒蟹」後便留下了深刻的印象，一直回味至今。辛辣的黑胡椒非但沒有蓋過大蟹的鮮味，反而更喚醒了味蕾，愈食愈惹味，最後就連一滴汁液都不能放過！

This is a fabulous dish. I once savoured this Black Pepper Crab in Singapore. This dish lingered in my mind. I was surprised that the hot black pepper not only did not over-power the freshness of the crab, it excited the taste buds instead. Even the sauce was good to the last drop!

材料

肉蟹	600 克
麵粉	1 湯匙
黑胡椒粒	1/4 杯（略壓碎）
蒜茸	2 湯匙
乾葱	4 粒（切片）
葱	2 條（切段）
紅辣椒	1 隻（切片）
蝦米辣椒醬	1 湯匙
九層塔	1 棵（取葉）
牛油	30 克

調味料

生抽	1 湯匙
蠔油	2 湯匙
糖	1 湯匙

做法：

1. 肉蟹沖淨及斬件，隔淨水分，撒下 1 湯匙麵粉拌勻。

2. 燒 6 湯匙油，放入蟹件半煎炸至八成熟，取出。

3. 剩 2 湯匙油加牛油爆香黑胡椒粒及蒜茸，加乾葱、葱、辣椒及蝦米辣椒醬炒香。

4. 蟹件回鑊，加調味料兜炒均勻，最後拌入九層塔。

Ingredients

600g crabs
1 tbsp flour
1/4 cup black peppercorn, lightly crushed
2 tbsp chopped garlic
4 shallots, sliced
2 stalks spring onion, sectioned
1 red chilli, sliced
1 tbsp chilli prawn paste
1 stalk Thai basil leaves
30g butter

Seasonings

1 tbsp light soy sauce
2 tbsp oyster sauce
1 tbsp sugar

Method

1. Rinse and dress crabs, chop into bite-size pieces, drain well. Dust 1 tbsp flour over crab pieces

2. Heat 6 tbsp oil, lower crab pieces into hot oil and fry until 80% cooked, remove and drain away oil leaving 2 tbsp.

3. Heat 2 tbsp oil and butter, sauté black peppercorn and garlic, add shallots, spring onion, red chilli and chilli prawn paste, fry until fragrant.

4. Add crab pieces and seasonings, stir-fry together. Lastly add Thai basil leaves.

Note

· 蟹件薄薄撲上麵粉可減少表面濕度，使蟹件較乾身，煎時更金黃香口。

· 九層塔是羅勒的一種，味道辛香，配海鮮、肉類可增添食物滋味。羅勒的品種繁多，廣泛應用於意大利、東南亞及台灣菜式中。

· Dusting flour over the crab pieces helps to absorb surface moisture, keeping them dry so that they become golden brown more readily.

· Thai basil with a sweet, strong and pungent taste is one of the many species of the basil family. It is widely used to accompany seafood and meat dishes. Other varieties also play a prominent role in Italian, Southeast Asian and Taiwanese cuisine.

鹽 Salt

鹽是所有鹹味菜式的根源。

在烹調中有食鹽、海鹽、岩鹽等等。鹽又可加工製成多種
調味品，例如醬油、蠔油、魚露、醬料及其他鹽醃漬食材。
鹽可說是「百味之王」。

Salt is the basic element to savoury dishes.

There are table salt, sea salt, rock salt, etc., commonly used in
cooking. It is also a key component in the making of soy sauce,
oyster sauce, fish sauce, other savoury sauces, pastes, as well as
preserves. It can be called 'the king of flavours'.

淮鹽
Spiced Salt

材料

鹽	2 湯匙
五香粉	1 平茶匙

做法

1. 乾鑊炒鹽至乾爽，離火，加五香粉拌匀。鹽的熱力已可把五香粉的香氣溢出，所以切勿炒五香粉，因乾粉會易炒燶。

2. 待涼，存放密封玻璃瓶內可保持新鮮數星期。

Ingredients

2 tbsp salt
1 level tsp five spiced powder

Method

1. Dry-fry salt until light and colour changes. Remove from heat and add five spiced powder. Toss together until well mixed. The hot salt will extract the fragrance from the powder. Do not fry the powder as it will burn easily.

2. Cool and store in air-tight glass container. Use them fresh within a few weeks.

麻辣椒鹽
Sichuan Pepper Salt

材料

四川紅花椒粒	1 湯匙
鹽	2 湯匙

做法

1. 乾鑊炒紅花椒粒至溢出香氣。取出磨碎。過篩，掉去梗及雜物。

2. 乾鑊炒鹽至乾爽，加紅花椒碎一起炒香。

3. 待涼，存放密封玻璃瓶內可保持新鮮兩星期。

Ingredients

1 tbsp Sichuan peppercorns
2 tbsp salt

Method

1. Dry-fry Sichuan peppercorns until fragrant. Remove and grind with pestle and mortar. Sieve to remove any stalks or impurities.

2. Dry-fry salt until light and colour changes, add ground Sichuan peppercorns and toss together.

3. Cool and store in air-tight glass container. Use them fresh within 2 weeks.

木魚紫菜鹽
Bonito Nori Salt

材料

鹽	1 湯匙
木魚紫菜口式飯調味料	1 湯匙

做法

1. 乾鑊炒鹽至乾爽。離火，加入調味料拌勻。切勿炒調味料因木魚紫菜已是乾身，加熱會易炒燶。

2. 待涼，存放密封玻璃瓶內可保持新鮮一星期。

Ingredients

1 tbsp salt
1 tbsp Japanese bonito nori seasonings for rice

Method

1. Dry-fry salt until light and colour changes. Remove from heat and add bonito nori seasoning. Mix together well. Do not fry the seasonings as they are light and dry and will burn easily.

2. Cool and store in air-tight glass container. Use them fresh within 1 week.

芝麻鹽
Sesame Salt

材料

白芝麻	2 湯匙
鹽	1/2 茶匙

做法

1. 乾鑊慢火炒白芝麻至微黃色，不停炒拌避免焦燶。加鹽一起炒勻。

2. 待涼，存放密封玻璃瓶內可保持新鮮兩星期。

Ingredients

2 tbsp white sesame seeds
1/2 tsp salt

Method

1. Dry-fry white sesame seeds in the wok, tossing the whole time to prevent burning. When sesame seeds begin to change to light golden, add salt and fry together.

2. Cool and store in air-tight glass container. Use them fresh within 2 weeks.

泰式鹽焗烏頭
Salt-baked Grey Mullet Thai Style

鹽除了是調味料外，它亦有良好的導熱功能，因此有很多菜式如古法鹽焗雞、鹽焗魚等的做法都是把食材用大量鹽掩蓋，然後再用高溫把鹽加熱，待鹽吸收了足夠熱力後把食物煮熟。這樣去烹調食物的好處是不用把食材直接暴露於高溫之下，水分因此得以保留，食材便可保持濕潤了。

Other than being a seasoning, salt can also retain heat and act as a cooking component. Dishes like "Authentic Salt-baked Chicken" and "Thai Style Salt-baked Fish" are all cooked with a generous amount of salt. The hot salt sears the food, seals in the moisture and keeps the meat tender and juicy.

材料

烏頭連鱗	1 條（重約 500-600 克）
幼鹽	1/2 茶匙
蛋白	1 個
粗鹽	12-15 湯匙

香料

香茅	1 枝（略拍）
南薑	4 片（略拍）
檸檬葉	4 塊（略撕開）
乾葱	2 粒（略拍）
紅辣椒	1-2 隻（略拍）

泰式酸辣汁

魚露	3 湯匙
青檸汁	3-4 湯匙
白醋	1 湯匙
糖	2-3 茶匙
青辣椒	1 隻
青指天椒	1 隻
蒜頭	1 粒

Ingredients

- 1 grey mullet with scales on (500-600g)
- 1/2 tsp salt
- 1 egg white
- 12-15 tbsp coarse salt

Herbs

- 1 lemongrass, lightly crushed
- 4 slices galangal, lightly crushed
- 4 kaffir lime leaves, lightly torn
- 2 shallots, lightly crushed
- 1-2 bird's eye chillies, lightly crushed

Thai spicy vinaigrette

- 3 tbsp fish sauce
- 3-4 tbsp Thai lime juice
- 1 tbsp white vinegar
- 2-3 tsp sugar
- 1 green chilli
- 1 green bird's eye chilli
- 1 clove garlic

香料如香茅、乾葱、辣椒等要用刀略拍才容易出味，至於檸檬葉宜用手撕開。

Lightly crush the lemongrass, shallots and red chilli with a chopper to release their fragrance. Tear apart the kaffir lime leaves for the same purpose.

做法

1. 原條烏頭連鱗劏肚取出內臟，沖淨及抹乾。用 1/2 茶匙鹽搓勻肚內，塞入香料。用蛋白掃勻魚身以便沾上粗鹽。

2. 錫紙上撒 4 至 5 湯匙粗鹽，撥平至烏頭長度。放上烏頭，再用粗鹽均勻蓋滿，覆起錫紙包裹整條魚；將魚轉至焗盆。

3. 放入預熱焗爐攝氏 200-220 度，焗約 20-25 分鐘（視乎魚大細）。若想魚肉甘香，20 分鐘後可小心打開錫紙，繼續焗 8 至 10 分鐘至面鹽呈乾身及微微金黃，取出上碟。

4. 食時用刀叉撥開魚皮及鹽，取出香料，趁熱用泰式酸辣汁蘸魚肉吃。

泰式酸辣汁做法

1. 舂碎或用刀剁碎青辣椒、青指天椒及蒜頭。

2. 加魚露、青檸汁、白醋及糖拌勻。試味，照個人喜愛調校。

Method

1. Slit open stomach of grey mullet with scales on, remove entrails, rinse and wipe dry. Rub the inside of the fish with 1/2 tsp salt. Stuff in herbs. Glaze whole fish with egg white.

2. Sprinkle 4 to 5 tbsp coarse salt on the foil, spread out to the length of the fish. Place fish on the salt, cover whole fish with more coarse salt. Wrap fish with foil. Transfer to the baking tray.

3. Bake fish in preheated oven 200-220°C for 20-25 minutes (depending on the size of the fish). If prefer the fish meat a little dryer, slit open the foil after 20 minutes, and continue baking for 8 to 10 more minutes until the salt appears dry and light golden colour. Remove to a plate.

4. Peel off fish skin and salt with knife and fork, discard herbs. Serve fish meat with Thai spicy vinaigrette.

Thai spicy vinaigrette

1. Place chillies and garlic in a pestle and mortar and pound to a rough paste. Or chop finely.

2. Add fish sauce, lime juice, vinegar and enough sugar to taste. Mix well and adjust taste according to your liking.

鹽烤螃蟹
Salt-baked Crabs

想要嘗蟹肉的真滋味，除了清蒸外，鹽烤亦可做到同樣效果。把螃蟹埋於炒香的鹽巴內，數分鐘後隨即飄來陣陣香氣。把鹽撥開，立即便見到橙紅紅的螃蟹，還有從蟹爪間滲出的蟹油……多滋味啊！

Steaming and salt-baking are the best ways to savour the freshness of the crab. Simply cover the crabs with fried hot salt, and within minutes, a subtle sweetness filled the air. Push the salt aside and you can see the crabs are cooked with a bright orange-red colour, and with oil sipping out from the legs too. It's luscious!

材料

螃蟹	4 隻（每隻約 150 克）
粗鹽	1.5 千克 -1.8 千克（視乎炒鑊大小而定）
八角	10 粒

Ingredients

4 live fresh water crabs (each weighs 150g approximately)
1.5kg – 1.8kg coarse salt (depending on the size of the wok)
10 star anise

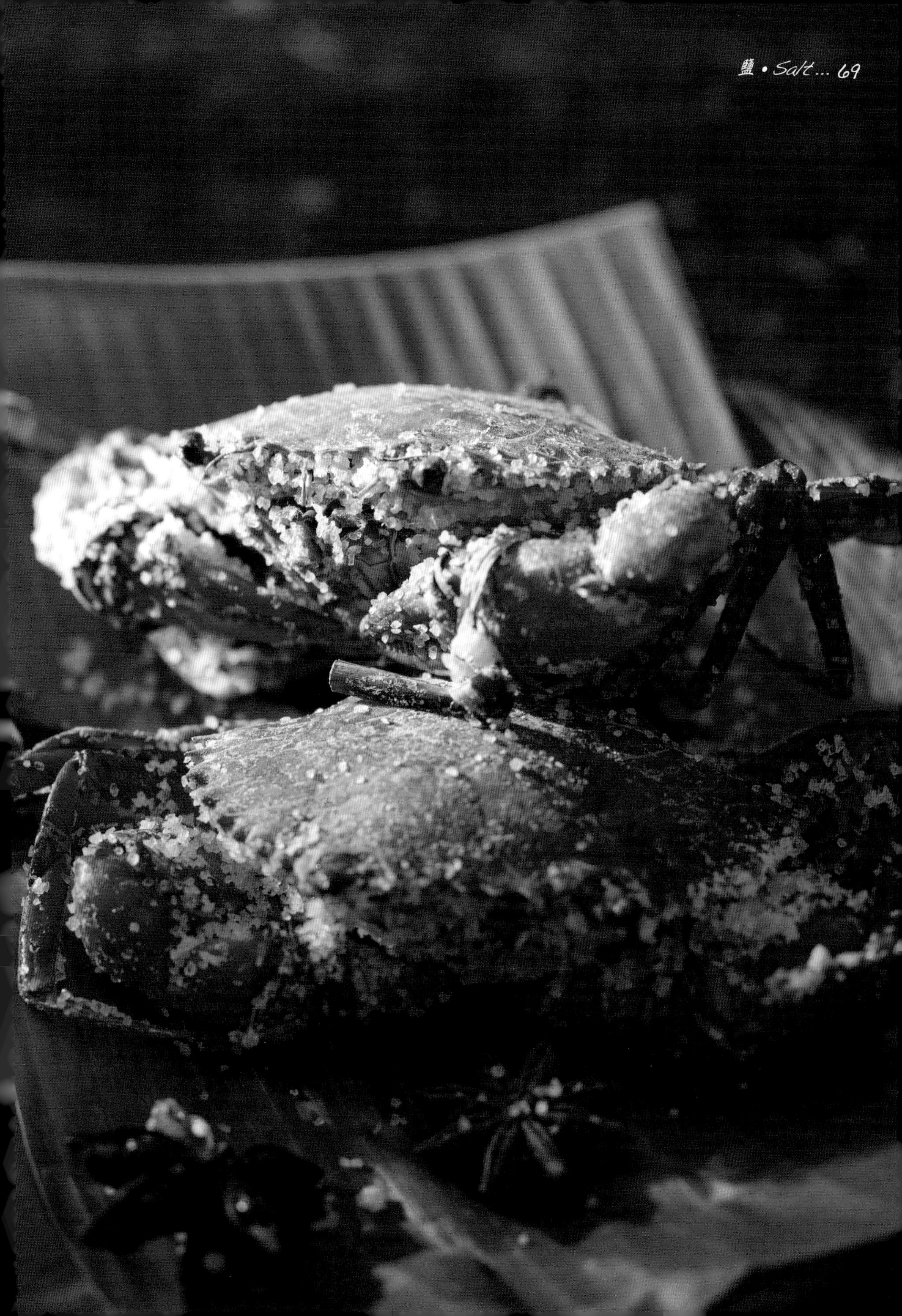

做法

1. 螃蟹擦乾淨，沖洗，隔乾。

2. 粗鹽與八角放乾鑊內炒香，炒至鹽大熱，然後取出三份之二。

3. 將鑊內剩下的鹽撥平，排放螃蟹（不可疊起），以蟹蓋向下。

4. 把取起的鹽放回，均勻蓋過每隻螃蟹。

5. 蓋上鑊蓋，以中小火保持溫度，烤焗 12 至 15 分鐘。（視乎螃蟹的大細）

6. 螃蟹烤熟後，撥開鹽，取出螃蟹，再用小掃清理螃蟹上的鹽巴，便可上碟趁熱品嘗。

Method

1. Rinse and scrub crabs clean. Drain well.

2. Place salt and star anise in a wok, fry over medium-high heat until hot and fragrant. Take out two-thirds of the hot salt.

3. Spread out the remaining one-third hot salt evenly in the wok. Arrange crabs upside down on the hot salt (do not stack them over one another).

4. Cover evenly with the rest of the salt, making sure the crabs are completely covered.

5. Put wok cover on and cook over medium-low heat for 12 to 15 minutes (depending on the size of the crabs), keeping the temperature constant the whole time.

6. Remove crabs when cooked, brush away the coarse salt and serve at once.

Note

· 先將三份一鹽與八角同炒，待八角出味後才倒入其餘鹽炒香。

· 要測試炒鹽是否夠熱，可灑少許水在鹽上，如立即蒸發成蒸氣，表示鹽已夠熱可焗蟹了。

· 將蟹蓋向下的原因是可避免蟹膏從腳部溢出，浪費美味。

· 把用完的鹽先翻炒至乾身，待涼後可儲起留用。

· Fry one-third of the salt with star anise until fragrant. This will extract the spicy aroma from the star anise. Then add the remaining salt and fry well together.

· Simply splash a few drops of water onto the hot salt to test its temperature. If steam emits from the salt, the temperature is high enough to cook the crabs.

· Place crabs upside down on the heated salt to prevent the delicious crab roe from sipping out of from the legs.

· Fry the used salt until dry, let cool and store for the next cooking.

新會渣鴨
Braised Duck with Ginger

此菜故名思義就是新會的地方菜,「渣」的意思未能深究,相信是燜燴慢煮的意思,是我向一位伯母屈老師偷師的。有一次在友人家嘗過此鴨,鴨味香濃,層次分明,我大為欣賞!隨即當然是向伯母請教,伯母亦很樂意一五一十的告知製法。想不到如此食味豐富的燜鴨,原來用的調味極簡單,就只是鹽而已!借此機會,再次感謝屈老師的分享,讓大家有機會嘗試一下鹽對食材味道改變的奧妙。

This is an authentic dish of Xinhui. I once tasted this dish prepared by my friend's mother. It was so delicious that I quickly asked for the recipe. To my surprise, salt was the only seasoning used. It is the salt, combined with slow cooking, that bring out the meaty flavour of the duck.

材料

光鴨	1 隻(約 1.8 千克)
薑汁	3 湯匙
薑(連皮)	200-250 克
蒜頭	8 粒
鹽	5-6 茶匙
油	1 杯
水	適量

Ingredients

1 dressed duck (approximately 1.8kg)

3 tbsp ginger juice

200-250g ginger with skin

8 cloves garlic

5-6 tsp salt

1 cup oil

water for braising

做法

1. 光鴨原隻洗淨及抹乾。用薑汁搽勻鴨肉，待至皮略乾身。

2. 薑連皮沖淨，切厚件略拍。

3. 用乾鑊炒鹽至乾爽，加油再炒勻，然後加薑及蒜頭炒香。取出薑及蒜，舀起約 2 茶匙鹽作蘸料。

4. 鴨放入油內，慢火煎至微黃色。濾出多餘油，薑蒜回鑊，加適量水蓋過半隻鴨身，蓋上鑊蓋用中慢火燜腍（約 1 至 1 1/4 小時，視乎鴨大小），每約 15 分鐘把鴨反轉，需要時加適量水避免燜乾水分。

5. 鴨燜熟及腍後取出斬件，排在碟上，淋上少許燜汁，再以炒過的鹽蘸吃。

Note:

· 煎鴨時防止濺油的方法，除了抹乾鴨身的水分外，也要刺穿鴨的眼睛，避免鴨眼爆開時弄傷你的肌膚。

· To avoid hot oil splashing on your hands, wipe duck dry, pierce through the eyes to prevent them from splitting.

Method

1. Use whole duck, rinse and wipe dry. Rub duck with ginger juice. Set aside until the skin is slightly dry.

2. Rinse ginger with skin, cut into thick slices and lightly crush.

3. In a dry wok, fry salt until dry, add oil and fry for a while. Add ginger and garlic and fry until fragrant. Remove ginger and garlic, set aside. Reserve 2 tsp of salt for dipping.

4. Lower duck into medium-hot oil and fry until the skin is evenly coloured. Drain off excess oil, return ginger and garlic to the wok, pour in enough water to cover half the duck, cover with the lid and braise over medium-low heat until duck is tender and cooked (approximately 1 to 1 1/4 hours depending on the size of the duck). Turn the duck every 15 minutes to ensure evenly cooked. If necessary, add a little more water to prevent drying.

5. When duck is cooked, remove and chop into bite-size pieces, arrange on a plate. Pour a little braising liquid over duck and serve with dipping salt.

椒鹽魷魚總有一股令人難以抗拒的魅力！它那酥脆的外層，略帶柔韌的魷魚片，一口咬下鮮味澎湃，加上椒鹽的辛香，讓人一件接一件，停不了口！

Crispy fried squid is always difficult to resist. The stringy texture of the squid and its crispy coating offer a delicate balance to every bite. The final touch with spiced salt heightens the flavour. How can one turn down a second helping!

材料

鮮魷魚	1-2 隻（500克）
蛋	1/2 個（拌勻）
生粉	1 杯
紅辣椒碎	1 茶匙
蒜茸	1 茶匙
淮鹽	1/2 茶匙

醃料

薑汁	1/2 湯匙
黃酒	1/2 湯匙
淮鹽	1/2 茶匙

Ingredients

1-2 squids (500g)
1/2 beaten egg
1 cup potato starch
1 tsp chopped red chilli
1 tsp chopped garlic
1/2 tsp spiced salt

Marinade

1/2 tbsp ginger juice
1/2 tbsp brown wine
1/2 tsp spiced salt

椒鹽鮮魷
Deep-fried Squid with Spiced Salt

做法

1. 鮮魷魚取出內臟，沖淨及抹乾，於魷魚內邊剐十字紋，再切件，與醃料拌勻醃5分鐘。

2. 蛋液與魷魚拌勻，沾上生粉，拍去多餘的粉。

3. 燒 1/3 鑊油，油熱時把魷魚逐件放入油內，用熱油炸至捲起成筒形，取出濾油。

4. 再把鑊內的油燒熱，魷魚回鑊炸至金黃香脆，取出隔淨油分。

5. 燒 1/2 湯匙油爆香紅辣椒及蒜茸，魷魚回鑊一起兜炒，多撒些淮鹽拌勻會更添惹味，上碟趁熱品嘗。

** 淮鹽做法請參照第 60 頁

Method

1. Clean and rinse squid, wipe dry. Slit open the body, mark criss-cross on the inner side, then cut into bite-size pieces. Mix with marinade and set aside for 5 minutes.

2. Mix squid pieces with beaten egg, then coat evenly with potato starch, shake off excess starch from squid.

3. Heat 1/3 wok of oil, lower squid pieces one by one into the oil and deep-fry in hot oil until they curl up. Remove to drain.

4. Reheat oil in the wok, deep-fry squid pieces a second time until golden brown and crispy. Take out and drain off excess oil.

5. Heat 1/2 tbsp oil in the wok, sauté red chilli and garlic until fragrant. Return squid pieces to the wok and toss together with enough spiced salt to taste. Serve at once.

**Spiced Salt refer to P.60

新派鹽焗雞
Home-style Salt-baked Chicken

擔任烹飪導師廿多年，但我仍非常享受這份工作，其中一個原因便是讓我常常思考怎樣可把酒樓大菜化繁為簡，讓學生們在家也可做到。例如這道鹽焗雞，傳統方法相對複雜，既要費力炒鹽，烹調時間亦較長，一般家廚都會敬而遠之。於是我便想到用焗爐幫手，先醃、蒸，然後再烤焗至皮乾身，食落皮香肉嫩，跟酒樓大師傅的手勢真可一較高下呢！

Even after twenty years as a cookery instructor, I still have the same great passion for the job. One of my challenges is to simplify restaurant style dishes for home cooking. This "Home-style Salt-baked Chicken" is one of these creations. The chicken is marinated, steamed and browned in the oven to achieve restaurant-like result.

材料

光雞	1 隻（1.2 千克）
薑	2 厚片（略拍）
葱	1 條（略拍）
八角	2 粒
生抽	適量（用來上色）
油	適量

醃料

薑汁	2 湯匙
鹽	4 茶匙
雞粉	1 茶匙
沙薑粉	1/2 茶匙

沙薑蘸汁

沙薑粉	2 湯匙
鹽	1/4 茶匙
油	1-2 湯匙

薑葱油

薑茸	2 湯匙
葱茸	2 湯匙
鹽	1/4 茶匙
糖	1/4 茶匙
油	2 湯匙

做法

1. 光雞洗淨，抹乾，用醃料搓勻全身，醃約 1 小時。

2. 燒滾水，光雞醃後瀝乾；把薑、葱及八角放雞腔內，雞盛碟上，蒸約 20-25 分鐘，離火，掉去薑葱，濾淨雞腔內水分。

3. 用生抽搓勻雞皮上色，再掃上油。雞放鐵架上，用鐵盆墊底。

4. 放入預熱焗爐攝氏 200 至 220 度，焗 10 至 15 分鐘至皮呈金黃香脆，取出略凍便可斬件上碟。

5. 沙薑蘸汁及薑葱油分別拌勻，與鹽焗雞伴食。

Ingredients

1 dressed chicken (1.2kg)
2 thick slices ginger, lightly crushed
1 stalk spring onion, lightly crushed
2 star anise
a little light soy sauce for skin colour
a little oil for brushing

Marinade

2 tbsp ginger juice
4 tsp salt
1 tsp chicken powder
1/2 tsp lesser galangal powder

Ginger flavoured dip

2 tbsp lesser galangal powder
1/4 tsp salt
1-2 tbsp oil

Spring onion ginger dip

2 tbsp grated ginger
2 tbsp minced spring onion
1/4 tsp salt
1/4 tsp sugar
2 tbsp oil

Note

· 如家中沒有焗爐，可用滾油淋雞皮，直至雞皮香脆。

· 先將雞蒸熟，然後才將雞用快火焗至金黃香脆，可令雞肉保留肉汁。

· If oven is not available, pour hot oil over the chicken repeatedly until the skin is crispy and brown.

· Steaming the chicken until almost cooked, before quick-browning the skin in the oven will retain its juiciness.

Method

1. Clean chicken and wipe dry, rub marinade over the skin and inside the chicken and set aside for 1 hour.

2. Bring half wok of water to a boil. Drain marinade off chicken, stuff ginger, spring onion and star anise into the cavity, place chicken on a plate and steam for approximately 20-25 minutes. Remove from heat, discard ginger, spring onion, star anise, and drain away any juice.

3. Rub soy sauce over the skin, brush evenly with oil. Place chicken on a rack over a drip tray.

4. In a preheated oven 200 to 220°C, roast chicken for 10-15 minutes until the skin is lightly brown. Remove to cool. Chop into bite-size pieces and arrange on a plate.

5. Mix ginger flavoured dip and spring onion ginger dip separately. Serve with salt-baked chicken.

麻油
Sesame Oil

亞洲料理使用的麻油（又叫香油）是一種很神奇的調味料，只需灑下數滴，便立即把整道菜的味道都提升了。特別是用於前菜冷盤的調味，更可讓食客充分品味麻油的幽香。

選購麻油時記緊要參考標籤上的成分，以 100% 芝麻製造的純麻油為佳，香味濃郁。市面上亦有些是芝麻油混合菜油，香味當然不及 100% 的純麻油，但售價相對較便宜。

In Asian cooking, sesame oil or fragrance oil is incredibly versatile. Just a few drops will enhance the flavour of any food. This is especially true for starters, as its subtle aroma will excite the palate of every diner.

Check the label on the bottle. Only pure sesame oil is made with 100% sesame seeds. Its counterpart, which is oil blended from sesame oil and vegetable oil, is less expensive and has inferior aroma and quality.

涼拌菠菜
Pickled Spinach

材料

菠菜	300 克
蒜茸	2 茶匙
凍飲用水	4 杯
鹽	1 茶匙
糖	1 茶匙
靚麻油	2-3 湯匙
鹽	適量（調味）

做法

1. 菠菜沖淨及整理好。

2. 燒半鑊水，加鹽及糖各 1 茶匙，放入菠菜，用中火煮熟（約 3-4 分鐘），取出，放入凍飲用水內過冷，隔淨。拌入蒜茸、適量鹽及麻油調味。放雪櫃內醃泡半天。

3. 食時隔淨汁液，把菠菜上碟成前菜。

Ingredients

300g spinach
2 tsp chopped garlic
4 cups cooled boiled water
1 tsp salt
1 tsp sugar
2-3 tbsp sesame oil
a little salt to taste

Method

1. Trim and rinse spinach.

2. Heat half wok water, add 1 tsp salt and 1 tsp sugar, cook spinach over medium heat for 3 to 4 minutes. Remove and refresh with cooled boiled water, drain well. Mix with chopped garlic, enough salt and sesame oil to taste. Chill in fridge for half a day.

3. To serve: Drain spinach to remove the liquid, transfer to a small plate and serve as an appetizer.

Note

· 菠菜不要灼得過腍，除了沒有嚼口外，也會令營養流失。

· Do not overcook the spinach or they will become mushy. Overcooking the spinach will also result in the loss of nutrients.

涼拌大豆芽
Pickled Soy Bean Sprouts

材料

大豆芽	300 克
韭菜	4 條
凍飲用水	4 杯
鹽	1 茶匙
靚麻油	2-3 湯匙
鹽	適量（調味）

Ingredients

- 300g soy bean sprouts
- 4 stalks green Chinese chive
- 4 cups cooled boiled water
- 1 tsp salt
- 2-3 tbsp sesame oil
- a little salt to taste

做法

1. 大豆芽沖淨及整理好。韭菜沖淨及切段。

2. 燒半鑊水，加 1 茶匙鹽，放入大豆芽，用中火煮 5-6 分鐘至黃豆腍，取出，放入凍飲用水內過冷，隔淨。再燒滾鑊內之水，把韭菜灼熟，取出，放入凍飲用水內過冷，隔淨。

3. 大豆芽與韭菜放大碗內，拌入適量鹽及麻油調味，放雪櫃內醃泡半天。

4. 食時隔淨汁液，把芽菜韭菜上碟成前菜。

Method

1. Trim and rinse soy bean sprouts. Rinse and cut green Chinese chives into sections.

2. Heat half wok water, add 1 tsp salt, cook soy bean sprouts for 5-6 minutes until tender, remove and refresh with cooled boiled water, drain well. Bring water to a boil again, parboiled green Chinese chives, remove and refresh with cooled boiled water, drain well.

3. Place soy bean sprouts and green Chinese chives in a mixing bowl, mix with enough salt and sesame oil to taste. Chill in fridge for half a day.

4. To serve: Drain soy bean sprouts and green Chinese chives to remove the liquid. Transfer to a small plate and serve as an appetizer.

Note

· 要保持大豆芽及韭菜的爽脆質感及鮮明色澤，在汆水撈起後最好立即放入凍水內沖泡降溫。為了保持衛生，應使用飲用凍水。

· To maintain the colour and crispiness of the soy bean sprouts and green Chinese chives, rinse them with cooled boiled water as a soon as they are removed from the heat. For hygiene purpose, be sure to rinse the cooked vegetables with cooled boiled water.

拍青瓜
Sesame Flavoured Cucumber

Note

· 青瓜放入保鮮袋內，才用菜刀拍鬆，可避免拍青瓜時汁液四濺。

· 青瓜被拍至纖維鬆開，較易吸收調味，但不可過量拍爛。

· Place cucumber in a polythene bag before crushing it with a chopper. This will prevent the juice from splashing all over the place.

· Crushing the cucumber softens its tissue so it can absorb the seasonings better. Be sure not to crush the cucumber completely.

材料

青瓜	2 條（500 克）
鹽	1/2-3/4 茶匙
麻油	1-2 湯匙

隨意調味

A 黑醋	1-2 茶匙
B 豆瓣醬	1-2 茶匙

做法

1. 洗擦青瓜外皮，切開一半，用菜刀拍鬆，再切成小件。放大碗內與鹽拌勻，放雪櫃待 30 分鐘。這時青瓜會滲出水分。

2. 食前把青瓜隔去汁液，與麻油拌勻，上碟成涼菜。

可隨意加入：

A. 1-2 茶匙黑醋（可用中國黑醋，或用意大利陳醋更佳）：醋可中和麻油的膩，也可幫助消化。

B. 1-2 茶匙豆瓣醬：豆瓣醬味道鹹香帶少許辣，可提升這道拍青瓜成惹味前菜。

Ingredients

2 cucumbers (500g)
1/2-3/4 tsp salt
1-2 tbsp sesame oil

Optional seasonings

a. 1-2 tsp dark vinegar or balsamic vinegar
b. 1-2 tsp hot bean paste

Method

1. Scrub cucumber well, cut in half lengthwise, 'bang bang' with the flat blade of the chopper to crush the cucumber slightly. Then cut into bite-size pieces. Place them in a mixing bowl, mix thoroughly with salt. Chill in fridge for 30 minutes to let the moisture drain out from the cucumber.

2. Just before serving, drain cucumber well, mix with sesame oil. Serve as an appetizer.

According to one's taste:

a. Add 1-2 tsp dark vinegar or balsamic vinegar to harmonize the sesame oil as well as help digestion.

b. Mix cucumber with 1-2 tsp hot bean paste to add a fiery note to this once mild appetizer.

南瓜毛豆仁
Pumpkin and Soy Bean Kernels

南瓜與毛豆，一軟一硬，口感味道卻出乎意料的配合。加上南瓜的
鮮黃及毛豆的翠綠配搭，單是賣相便已先下一城。做法亦毋須複雜，
只需用鹽水煮熟，撈起與少許麻油拌勻，放入雪櫃待冷凍後便成一
佐酒小吃。

Pumpkin and soy bean kernels have contrasting textures, yet they
go very well together. Furthermore, the yellow pumpkin and the jade
green bean kernels are visually stunning. Simply parboil them in salt
water, blend with sesame oil and chill to serve. They go very well
with wine.

材料

南瓜	300 克
急凍毛豆仁	1/2 杯
凍飲用水	4 杯
鹽	1 茶匙
鹽	1/2-3/4 茶匙
麻油	1-2 湯匙

做法

1. 南瓜去皮及瓤，沖淨切粒。

2. 燒半鑊水，加 1 茶匙鹽，放下南瓜粒飛水至剛熟（約 2-3 分
   鐘），取出，放入凍飲用水內過冷，隔淨。

3. 急凍毛豆仁解凍，沖淨，放入以上滾水內飛水，取出，放入凍
   飲用水內過冷，隔淨。

4. 把南瓜及毛豆仁放大碗內，加適量鹽及麻油拌勻調味，放雪櫃
   醃 30 分鐘即可享用。

Ingredients

300g pumpkin
1/2 cup frozen soy bean kernels
4 cups cooled boiled water
1 tsp salt
1/2-3/4 tsp salt
1-2 tbsp sesame oil

Method

1. Peel and remove seeds from pumpkin, rinse and cut into cubes.

2. Bring half wok water to a boil together with 1 tsp salt, lower pumpkin cubes into boiling water, cook until al dente (approximately 2 to 3 minutes), remove and refresh with cooled boiled water, drain well.

3. Thaw frozen soy bean kernels, blanch in the water, remove and refresh with cooled boiled water, drain well.

4. Place pumpkin cubes and soy bean kernels in a mixing bowl, season with salt and sesame oil to taste, mix well. Chill in fridge for 30 minutes before serving.

Note

· 市場上有不同產地的南瓜，又甜又粉的有日本南瓜，西菜用來焗或烤的有 Butternut Squash 南瓜。我就喜歡選用中國產啡色長身那種，價錢實惠，肉質 爽甜。汆水時最緊要是小心不要過熟，撈起後用凍飲用水過冷河，可保持質感。

· 除了買急凍毛豆仁外，也可以買連莢的新鮮毛豆，自己剝豆莢、挑豆。購買時 以豆莢青綠、有絨毛、飽滿為佳。

· Among the many different kinds of pumpkins in the marketplace, I prefer the reasonably priced Chinese long brown pumpkin, which has a sweet and slightly crunchy texture. Do not overcook them. Plunge them in cooled boiled water once cooked to retain its texture.

· Besides frozen soy bean kernels, there are also fresh ones in the pods available in the market. Choose soy beans with tiny silver hairs on the shell and with kernels bulging out.

芫茜荷蘭豆拌鮮腐竹
Coriander and Snow Peas with Fresh Soy Stick

素前菜的材料配搭很隨意，可隨你的個人口味去選配。我會以口感、香味及色澤來選擇用料，畢竟前菜的作用是要引起食慾，是一餐的開始！

Mix and match is my idea for a vegetarian starter. Bear in mind the texture, fragrance and colour when you choose the ingredients. This is important because a starter sets the tone for the whole meal!

材料

鮮腐竹	200 克
荷蘭豆	50 克
雲耳	6 朵（浸透）
芫茜	1 小束
白芝麻	1 茶匙
鹽	1 茶匙
靚麻油	2-3 湯匙
鹽	適量（調味）

做法

1. 鮮腐竹沖淨，切段；雲耳浸透後剪去硬端，然後撕碎；荷蘭豆撕去兩邊硬筋；芫茜沖淨及略切碎；白芝麻用白鑊炒香。

2. 燒半鑊水，加 1 茶匙鹽，先把鮮腐竹飛水（半分鐘），取出濾乾，再把雲耳飛水。最後把荷蘭豆飛水，取出用凍飲用水沖淨，然後切絲。

3. 全部材料放大碗內，加入麻油及適量鹽拌勻調味，撒下炒香白芝麻，拌一拌便成前菜。

Ingredients

200g fresh soy stick
50g snow peas
6 cloud ear fungus, soaked
1 small bunch coriander
1 tsp white sesame seeds
1 tsp salt
2-3 tbsp sesame oil
a little salt to taste

Method

1. Rinse fresh soy stick, cut into sections. Trim off the hard end of the soaked cloud ear fungus, tear into little pieces. Remove the strings on both sides of the snow peas. Rinse and roughly chop the coriander. Dry fry sesame seeds until golden in colour.

2. Bring half wok water to a boil, add 1 tsp salt, blanch fresh soy sticks for 1/2 minute, remove and drain well. Blanch cloud ear fungus and drain. Blanch snow peas, remove and refresh with cooled boiled water, drain well and cut into julienne.

3. Place all the ingredients in a mixing bowl, season with enough salt and sesame oil to taste. Mix with fried sesame seeds and serve as appetizer.

Note

· 無論是急凍或鮮腐竹必須要煮過。用飛水的方法把鮮腐竹以短時間煮一煮，既可使腐竹軟身，也可安全食用。切勿煮過火，鮮腐竹會溶化水中變成豆漿！

· Both frozen and fresh soy sticks must be cooked before serving. Blanch them for a short while to soften and to make them more hygienic to eat. Do not overcook, or they will dissolve in the water and become soy bean milk!

醬油
Soy Sauce

醬油是中菜烹調的調味磐石，它獨特的豉香味帶來了中菜的特質，絕對是不可或缺！

Soy sauce is the foundation stone of most Chinese dishes. Its unique flavour is the essence of Chinese cuisines. It is simply a must in the kitchen!

蒸魚豉油
Soy Sauce for Steamed Fish

材料

生抽	3 湯匙
水	1/3 杯
糖	1/2-3/4 茶匙
八角	1 粒
芫茜莖及根部 1-2 棵	

做法

1. 生抽、水、八角及芫茜放小鍋內煮滾，慢火煮 5 分鐘至味濃香。取出八角及芫茜，拌入適量糖調味。

2. 可趁熱淋於剛蒸好的魚上，或待涼放雪櫃，日後可隨時可用。

Ingredients

3 tbsp light soy sauce
1/3 cup water
1/2-3/4 tsp sugar
1 star anise
1-2 stalks and roots of coriander

Method

1. Place light soy sauce, water, star anise and coriander in a small saucepan and bring to a boil. Lower heat and boil for 5 minutes, until the fragrance of star anise and coriander has infused into the sauce. Remove star anise and coriander. Add sugar to complete the sauce.

2. Pour over steamed fish at once. Or store in fridge for later use.

Note

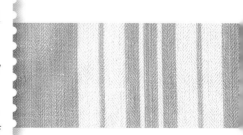

· 八角可提升生抽香味。
· 芫茜莖及根部一般都會棄掉。其實它的味道清香，洗淨後與生抽等一起煮，讓豉油的味道更佳。

· The star anise will enhance the flavour of the soy sauce with a touch of spice.
· Coriander stalks and roots are normally discarded. Rinse well and cook with soy sauce. These parts of the plant can impart fragrance to the sauce.

辣椒豉油
Chilli Soy Sauce

材料

醬油　　　　1/4 杯
（生抽與老抽各半取其色）
切碎指天椒　3-5 隻（視乎個人口味）

做法

把切碎的指天椒碎放醬油內浸泡至出味。

Ingredients

1/4 cup soy sauce (mixture of half light and dark soy sauce for colour)
3-5 chopped red bird's eye chillies (depending on individual taste)

Method

Place chopped red bird's eye chillies into the soy sauce and set aside to infuse flavour.

Note

· 即泡即食的辣椒豉油芳香撲鼻，若預先泡製，豉油只剩下辣味，香味遞減。

· Freshly made Chilli Soy Sauce has a fresh fragrance. If prepared way in advance, it will lose its fragrance and only taste hot.

古法豉油雞
Soy Sauce Chicken in Clay Pot

時代巨輪不停轉，社會的步伐愈走愈快，有時連
一些好東西亦在不知不覺間被淘汰了，實在可
惜！就好像瓦罉這個烹調好工具，其製造物料既
耐熱又傳味，只要調校至適當火候，毋須大火，
加點耐性，便可將食物「扣」得色、香、味俱
存，實非其他鍋具可代替的。只可惜現代人事事
求快，懂得欣賞它好處的人都不多了。在此僅呼
籲大家不要摒棄瓦罉，試試欣賞它的好處吧！

In the fast pace world of today, a lot of old
style goodies have slowly been forgotten.
The authentic clay pot is one of them. This
versatile cooking equipment is made with
material that retains both heat and flavour.
When cooked with the proper amount
of heat, the food will be slow-cooked to
perfection. In my opinion, there is no modern
cooking equipment that can replace the clay
pot. I sincerely hope that the clay pot will
continue to win back the hearts of cooking
enthusiasts.

材料

光雞	1 隻（1.2 千克）
薑	4 片
蔥	2 條

汁料

生抽	1/3 杯
老抽	3 湯匙
片糖	3/4-1 塊（剁碎）
水	2-2 1/2 杯
黃酒	3 湯匙

做法

1. 光雞沖淨及抹乾。

2. 瓦罉燒 6 湯匙油，爆香薑及蔥，放入光雞，用中火走油至雞皮呈金黃色，取出多餘油分。

3. 加生抽及老抽，邊煮邊淋至雞皮上色。加糖、黃酒及適量水至浸過半隻雞，煮滾。蓋上瓦罉蓋用中慢火煮10分鐘。把雞反轉至另一面，淋勻豉油，再蓋好煮 5 分鐘，重複兩至三次，視乎雞的大細，煮至雞熟及汁濃稠。

4. 取出雞，待略涼便斬件上碟。

5. 豉油汁試味，淋適量於雞件上。

Ingredients

1 chicken (1.2kg)
4 slices ginger
2 stalks spring onion

Sauce

1/3 cup light soy sauce
3 tbsp dark soy sauce
3/4-1 piece slab sugar, chopped
2-2 1/2 cups water
3 tbsp brown wine

Method

1. Rinse and wipe dry chicken.

2. In a clay pot, heat 6 tbsp oil, sauté ginger and spring onion, lower chicken into the oil and fry until all sides of the chicken are golden brown. Drain off excess oil.

3. Add light and dark soy sauce, baste over the chicken until evenly coloured. Add sugar, brown wine and enough water to cover half the chicken. Bring to a boil, cover and simmer over medium-low heat for 10 minutes. Turn chicken over to the other side, baste again, cover and cook for 5 minutes. Repeat this process two to three times (depending on the size of the chicken) until the chicken is cooked and sauce has thickened.

4. Remove chicken from sauce. Chop into bite-size pieces when it has cooled down a little. Arrange on a plate.

5. Adjust taste of the sauce, pour a little over the chopped chicken and serve.

瑞士雞翼
Chicken Wings in Swiss Sauce

瑞士雞翼蜚聲國際，稱得上是港式西餐的表表者，流行數十載，方興未艾。對於此菜的出處，又是否真的由瑞士傳入呢？要製作瑞士雞翼，主料是豉油、香料及冰糖。難道遠在他方的瑞士亦流行使用豉油烹調？翻查資料，其實是多年前有一位外籍遊客嘗過此菜後很喜歡，問道菜名，服務員答曰：Sweet Chicken Wings，遊客誤解為 "Swiss Chicken Wings"，自此其名不脛而走，造就了一道名菜的誕生。

This is a signature dish of Hong Kong style western cooking. In reality, this dish has nothing to do with Swiss at all. Only soy sauce, spices and rock sugar are used to cook the chicken wings. The story began with a foreign tourist asking the waiter for the name of a dish of chicken wings cooked in a sweet sauce, which he enjoyed very much. The waiter replied "Sweet Chicken Wings", which the tourist misinterpreted as "Swiss Chicken Wings". Henceforth, this dish has become very popular among locals and foreigners alike.

材料

雞全翼	8 隻

瑞士汁

水	4 杯
生抽	1/2 杯
老抽	1/2 杯
冰糖	1/2-2/3 杯
八角	2-3 粒
甘草	2-3 片
玫瑰露酒	1/2 湯匙

做法

1. 雞翼飛水，沖淨，隔乾。

2. 瑞士汁放鍋內煮滾，用中慢火熬 20 分鐘至味道香濃偏甜；試味。

3. 把雞翼放入瑞士汁內，先煮滾，隨即改用中慢火煮 5-8 分鐘，熄火浸 20 分鐘；取出上碟。

Ingredients

8 whole chicken wings

Swiss Sauce

4 cups water
1/2 cup light soy sauce
1/2 cup dark soy sauce
1/2-2/3 cup rock sugar
2-3 star anise
2-3 slices liquorice
1/2 tbsp Mei Kwei Lu Chiew

Method

1. Blanch and rinse chicken wings, drain well.

2. Place sauce mixture in a saucepan, simmer at medium heat for 20 minutes until fragrant and the taste is rich and sweet. Adjust taste.

3. Lower blanched chicken wings into sauce, bring back to a boil, reduce heat to medium-low and cook chicken wings for 5-8 minutes. Switch off heat, leave chicken wings in sauce for 20 more minutes. Take out chicken wings and serve.

Note

· 雞翼最適宜用中慢火燜熟，然後留在汁內浸至入味。
· 如果瑞士雞翼凍後呈現皺皮，這表示煮雞翼時用了猛火，或煮得過久。
· 這個甜豉油味道十分香濃。存放雪櫃，留作其他調味用途。

· Cook chicken wings over medium-low heat until just cooked and leave them in the sauce to absorb the flavour.
· If the skin of the chicken wings appears wrinkled, you either used high heat instead of medium-low, or cooked the wings a bit too long.
· This special sauce is extremely flavoursome. Store it in the fridge as seasoning for other dishes.

瑞士汁炒牛河

Stir-fried Rice Noodles with Sliced Beef in Swiss Sauce

由瑞士雞翼衍生出另一港式西餐名菜—瑞士汁炒牛河。乾炒牛河固然是廣東菜炒粉麵類中不能或缺的名菜,隨着瑞士雞翼的流行,廚師亦想到不如加入瑞士汁炒牛河,因而製成一道我認為略帶少許東南亞炒貴刁風味,但又不帶辛辣的甜味炒牛河。

This is a typical Hong Kong style dish which is derived from "Swiss Chicken Wings". The 'special sauce' is used to flavour the rice noodles instead of soy sauce. This adds a touch of sweetness to the dish and has some similarity to the Southeast Asian "Char Kway Teow", only without the spicy hot taste.

材料

炒河粉	400 克
牛肉	100 克
洋葱	1/4 個
葱	1 條(切段)
瑞士汁(燜瑞士雞翼的醬汁)	3 湯匙
銀芽	50 克

河粉要逐條分開。
Loosely separate the rice noodles before frying.

醃料

(一)

鬆肉粉或梳打食粉	1/4 茶匙
水	1 湯匙

(二)

蒜茸	1 茶匙
生抽	1 1/2 茶匙
糖	1/4 茶匙
生粉	1/2 茶匙
油	1 湯匙

做法

1. 河粉逐條分開放碟上,放入雪櫃或待至略乾身。

2. 牛肉橫紋切薄片,先與醃料(一)拌勻,再拌入(二)待 30 分鐘。洋葱切幼條;銀芽沖淨及隔乾。

3. 燒 1-2 湯匙油,先炒牛肉至九成熟,取出。

4. 再燒 2 湯匙油,把熱油盪勻鑊內,放入河粉輕輕弄散,兜炒至微黃,加入洋葱及葱炒香。倒入牛肉炒勻,邊炒邊加瑞士汁,炒至均勻便加銀芽,以猛火快手炒透,隨即上碟。

Ingredients

400g flat rice noodles for frying
100g beef flank
1/4 onion
1 stalk spring onion, sectioned
3 tbsp Swiss sauce
50g silver sprouts

Marinade (beef)

(A)
1/4 tsp meat tenderizer or cooking soda
1 tbsp water

(B)
1 tsp chopped garlic
1 1/2 tsp light soy sauce
1/4 tsp sugar
1/2 tsp potato starch
1 tbsp oil

Method

1. Loosen flat rice noodles and spread out on a plate to air-dry slightly, or dry in fridge.

2. Slice beef flank thinly across the grains. Mix with marinade (A) until well absorbed, add (B), mix well and set aside for 30 minutes. Cut onion into strips. Rinse silver sprouts and drain well.

3. Heat 1 to 2 tbsp oil, sauté sliced beef until almost cooked, set aside.

4. Heat 2 tbsp oil, swirl round sides of the wok, lower flat rice noodles into oil, gently loosen and stir-fry until well heated. Add onion and spring onion, toss together. Add fried beef slices and gradually add Swiss sauce to mix. Add silver sprouts. Stir fry quickly in high heat and serve immediately.

Note

· 將河粉放入雪櫃或待至略乾身才炒，待淋上汁料時，河粉可迅速吸收汁料的味道。

· Leave rice noodles on a plate in the fridge until slightly dry before stir-frying to prevent them sticking to the wok. The dryer the noodles, the easier they will absorb the sauce.

迷你東坡肉
Mini Dong Po Rou – Braised Pork

東坡肉膾炙人口，盛行多個世紀，至今仍大受歡迎。現代人雖云要避吃肥膩食品，但只要東坡肉一出現，還是抗拒不了！東坡肉究竟緣何有此魅力呢？或許就如蘇東坡自己所言：「慢火煮，少著水，火候足時它自美」。選材適當，簡單煮法，注意不要多用水，加上火候足夠，便成就了一道傳頌多個世紀的名菜。

"Dong Po Rou" has been a popular dish for centuries. Though most people prefer a low fat diet these days, this dish is still tough to resist! Why is this pork stew so irresistible? According to its creator Su Dong Po "slow cooking in minimum amount of liquid makes the pork flavoursome and tender".

材料

五花腩連皮	1 件（約 600-700 克重，15 厘米方形）
薑	2-3 厚片（略拍）
葱	1 條（略拍）
黃酒	2 湯匙
鹹水草	4 條（用熱水浸軟）

滷汁料

老抽	3-4 湯匙
生抽	2-3 湯匙
冰糖	3-4 湯匙
八角	3-4 粒
薑	4 片（略拍）
葱	1 條
黃酒	1/4 杯
水	約 2 杯（要浸過五花腩）

做法

1. 五花腩去淨毛，沖淨，把五花腩分切為 4 件方形小腩肉，用鹹水草紮好固定形狀。

2. 燒滾半鍋水，加入薑、葱及黃酒，放下五花腩件，出水 5 分鐘，取出。

3. 五花腩件放鍋內，加蓋面水及滷水料煮滾，改用慢火，蓋上鍋蓋，燜約 1 小時至腍。

4. 最後將汁煮至濃稠，與迷你東坡肉一起品嘗。

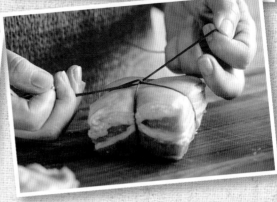

Ingredients

1 piece pork belly with skin (600-700g,15cm square)

2-3 thick slices ginger, slightly crushed

1 stalk spring onion, slightly crushed

2 tbsp brown wine

4 pieces straw, blanched until soft

Braising soy sauce mixture

3-4 tbsp dark soy sauce

2-3 tbsp light soy sauce

3-4 tbsp rock sugar

3-4 star anise

4 slices ginger, slightly crushed

1 stalk spring onion

1/4 cup brown wine

2 cups water, enough to cover pork belly

Method

1. Scrape clean skin of pork, rinse, cut into 4 square pieces. Tie a cross with straw on each piece to secure it's shape.

2. Blanch in half pot of boiling water together with ginger, spring onion and brown wine for 5 minutes, remove.

3. Arrange pork pieces in a pot, add enough water to just cover, add braising ingredients and bring to a boil. Reduce heat, cover and braise for 1 hour until tender.

4. Lastly, reduce sauce to gravy and serve with mini braised pork.

雙葱爆肥牛片
Sauté Beef with Two Onions

用「爆」來形容這道菜的烹調手法最適合不過。因為要做到牛肉香口及生熟適中，以及雙葱仍保留爽脆，必需要用熱鑊加快炒才可做到。預備功夫亦絕不可怠慢，否則耽誤了烹調時間便做不成應有的效果了。

"Sautéing" is the best way to cook this dish. For the onions to be crispy, and the beef to be tender and have a beefy flavour, they must be sautéed quickly and in high heat. To accomplish this, all the preparation work has to be done well in advance, otherwise it will be difficult to avoid overcooking this dish.

材料

肥牛肉片	250 克
洋葱	1/2 個
葱	4 條

醃料

生抽	1/2 湯匙
油	1 湯匙

芡汁

水	4 湯匙
老抽	1 湯匙
蠔油	1 湯匙
糖	1 湯匙
生粉	1 茶匙

做法

1. 肥牛肉片與醃料拌勻醃 5 分鐘。

2. 洋葱切條；葱切 5 厘米段，葱白略拍。

3. 燒熱 2 湯匙油，盪勻鑊內，放下牛肉，弄散，牛肉每邊烘一烘，炒至八至九成熟，取出。

4. 剩餘油爆炒洋葱及葱，牛肉回鑊兜炒，邊炒邊加入芡汁，猛火兜炒至芡汁略為收乾，上碟。

Ingredients

250g teriyaki beef slices
1/2 onion
4 stalks spring onion

Marinade

1/2 tbsp light soy sauce
1 tbsp oil

Sauce

4 tbsp water
1 tbsp dark soy sauce
1 tbsp oyster sauce
1 tbsp sugar
1 tsp potato starch

Method

1. Mix beef with marinade, set aside for 5 minutes.

2. Cut onion into strips. Cut spring onion into 5cm sections, lightly crush the white part.

3. Heat 2 tbsp oil, swirl round the sides of the wok, lower beef into the hot oil, loosen with chopsticks, fry each side until light golden brown and about medium cooked. Remove.

4. Sauté the two onions in the remaining oil. Return beef to the onions, toss well while adding sauce gradually. Lastly, fry everything together over high heat until sauce is reduced to a glaze. Ready to serve.

豉油皇薑葱煎雞
Pan-fried Chicken with Ginger and Spring Onion

冰鮮雞已成為香港人日常食材之一，經濟又方便。有人總以為冰鮮雞味道較活雞遜色，其實不然。簡單的配料、醬油與糖，就可煮出惹味香口的小菜。

Chilled chicken has become very popular as it is inexpensive and easy to get. But there are people who still prefer live chicken for its taste. In reality, simple seasonings like soy sauce and sugar can turn any type of chicken into the irresistible "Pan-fried Chicken".

材料

冰鮮雞	1/2 隻
薑	6 片（略拍）
葱	4 條（切 5 厘米段）

醃料

生抽	1 湯匙
生粉	1 茶匙
糖	1/2 茶匙
麻油及胡椒粉	少許
油	1 湯匙

芡汁

水	2 湯匙
生抽	2 茶匙
老抽	1 茶匙
糖	1 茶匙

做法

1. 雞沖淨抹乾，斬成小件，與醃料拌勻醃 10 分鐘。

2. 燒熱 3-4 湯匙油，放下雞件，用中慢火煎至兩面甘香，約 8 成熟，取出。

3. 剩 1-2 湯匙油爆香薑及葱，加芡汁煮滾至濃稠。

4. 煎香雞件放入芡汁內，用中猛火兜炒雞件至乾身即可。

Ingredients

1/2 chilled chicken
6 slices ginger, lightly crushed
4 stalks spring onion, cut into 5cm sections

Marinade

1 tbsp light soy sauce
1 tsp potato starch
1/2 tsp sugar
a little sesame oil and pepper
1 tbsp oil

Sauce

2 tbsp water
2 tsp light soy sauce
1 tsp dark soy sauce
1 tsp sugar

Method

1. Rinse and wipe dry chicken, chop into bite-size pieces. Mix with marinade and set aside for 10 minutes.

2. Heat 3-4 tbsp oil in the wok, lower chicken pieces into medium-low hot oil, shallow-fry each side until 80% cooked. Remove.

3. Drain away oil leaving 1-2 tbsp, sauté ginger and spring onion until fragrant. Add sauce and bring to a boil, simmer until gravy like.

4. Return fried chicken pieces to sauce, toss together over medium heat until chicken pieces are well glazed.

Note

· 可選擇半隻雞或淨肉的雞扒。煎香連骨雞件較淨肉口感層次多，味道更為甘香。

· Either half chicken or chicken fillet can be used. I prefer chicken with bones as it is more flavoursome when fried.

醬蘿蔔
Homemade Pickled Turnip

當我嘗第一口時，那清甜爽脆的獨特口感，讓我愛上了上海醬蘿蔔，至今仍回味無窮。所以每年到白蘿蔔當造的季節，我都會醃泡上十餘公斤，存放在玻璃瓶內，送禮自奉都適宜。識貨的好友們，亦會適時出現，以免錯過了好吃的醬蘿蔔！

From the first bite, I fell in love with this Shanghai style crunchy pickled turnip. Since then, when turnip is in season, I will pickle some ten kilograms of them, and store them in glass jars for myself and for my friends. Each year, these friends will always remember when to come by to pick them up, so they don't miss out on this delicious dish.

材料

白蘿蔔	600 克
鹽	1 平湯匙

滷汁

生抽	1/4 杯
砂糖	100-120 克
豆瓣醬	1 湯匙（或隨意）

做法

1. 蘿蔔去皮沖淨，切粒或條或片，放大碗內，灑下鹽拌勻，待最少半小時至汁液溢出。

2. 用潔淨毛巾吸乾蘿蔔水分。

3. 蘿蔔與滷汁拌勻，蓋好放入雪櫃醃最少兩日便可品嘗。

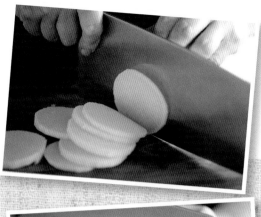

Ingredients

600g turnip
1 level tbsp salt

Pickling sauce

1/4 cup light soy sauce
100-120g sugar
1 tbsp hot bean paste, optional

Method

1. Peel turnip, rinse and wipe dry, cut into cubes or strips or slices, place in a mixing bowl, mix with salt and set aside for at least 30 minutes until juice drains out.

2. Squeeze out excess juice with a clean towel.

3. Mix turnip with pickling sauce, cover and store in fridge for at least 2 days before serving.

Note

· 白蘿蔔拌鹽醃泡片刻可迫出汁液，減低辛辣味。擠乾水分後的白蘿蔔會呈半透明，這會令白蘿蔔更易吸收滷汁。

· Marinate turnip pieces with salt until they turn opaque. This helps to soften the texture, as well as let out the excess moisture and sharp taste. Squeeze out the excess water from the turnip so it can better absorb the pickling sauce.

糖
Sugar

糖亦是重要的調味料，除了製作甜點，鹹、酸、苦、辣的味道都需要加入適量的糖提鮮及平衡，缺了它便會令菜式缺少了一個層次。

Sugar is another important fundamental seasoning. Apart from being the key ingredient in sweet dishes, it also enhances the flavour of other dishes, such as salty, sour, bitter and chilli-hot dishes. It adds a new dimension to these dishes.

拔絲蘋果
Crunchy Spun Sugar Apples

何謂拔絲蘋果？其實即是把白糖煮成焦糖後，再用炸香的蔬果滾上成脆糖面，是一道很受中外食家歡迎的經典甜點。蔬果選材的範圍很廣，如蘋果、香蕉、番薯、南瓜等都適宜。做法是先把蔬果塊沾上炸漿炸香，然後煮焦糖，把蔬果塊放入焦糖內拌勻。期間焦糖會拔起成絲，因而得名。臨進食前把蔬果塊放入冰水內一浸，焦糖漿隨即遇冷變硬，便成脆皮。這道甜點雖然頗考工夫，但效果良好。你有信心挑戰嗎？

What are Crunchy Spun Sugar Apples? They are apple fritters with a crispy caramel coating. Crunchy Spun Sugar Apples are very popular in both Eastern and Western cultures. You can use apple, banana, sweet potato or pumpkin for the fritter. Simply coat the ingredient with batter, deep-fry, then glaze with caramel. When lift with chopsticks, the syrup draws a long thread just like spun sugar. This is how it gets the name Spun Sugar. Dip them at once into ice water and they will harden and become brittle. It is a bit tricky, but worth trying!

材料 / *Ingredients*

蘋果	1 個	1 apple
炒香白芝麻	1 茶匙	1 tsp fried sesame seeds
冰水	1 大碗	1 mixing bowl of ice water

炸漿 / *Batter*

麵粉	100 克	100g flour
鹽	1/8 茶匙	1/8 tsp salt
水	適量（約 1/3 杯）	enough water to mix, approximately 1/3 cup

糖膠 / *Syrup*

白砂糖	120 克	120g sugar
水	1/8 杯	1/8 cup water

做法

1. 炸漿材料放大碗內,拌勻至滴珠程度。

2. 蘋果去皮及芯,切角。撒上 1-2 湯匙麵粉,使蘋果略為乾身,隨即放炸漿內避免變色。

3. 燒熱 1/3 鑊油,把沾滿炸漿的蘋果逐件放入油內,用中火炸至金黃香脆,取出,瀝淨油分。

4. 鑊內剩 1 湯匙油,沿鑊周邊盪勻,加糖及水煮滾,中火煮至糖膠狀,糖膠呈微黃色時,便把炸好的蘋果放入糖膠內拌勻,讓糖膠均勻裹着每件蘋果。這時糖膠的黏性變強,每當用筷子拔起一件蘋果時,糖膠即呈絲狀。

5 上碟前撒下炒香白芝麻。

6. 食時把每件蘋果浸一浸冰水,使糖膠凝固。

Method

1. Mix batter ingredients to a dropping consistency.

2. Peel and core apple. Cut into bite-size chunks. Sprinkle apple chunks with 1-2 tbsp flour and add to batter to prevent discolouring.

3. Heat 1/3 wok oil. Fry apple pieces in medium heat until golden and crispy. Remove and drain.

4. Leave 1 tbsp oil in the wok, swirl to all sides, add in the syrup ingredients of water and sugar. Boil over medium heat until syrup starts to caramelize. Add fried apple at once and toss until each piece is evenly coated with caramel. At this time, the caramel begins to change its consistency. When lift with chopsticks, the syrup draws a long thread like spun sugar.

5. Sprinkle with fried sesame seeds.

6. Dip briskly into ice water to harden the caramel. Serve at once.

Note

· 糖溶後不可攪拌，以免糖轉變成結晶粒。
· 測試糖膠是否可拔絲？可將呈微黃色的糖膠滴進冰水內，如變硬、變脆即表示糖膠可用來拔絲了。
· 冰水可助降低焦糖的溫度，即時變成硬脆。

· Do not stir the syrup after sugar has completely dissolved, or the syrup will crystalize.
· Testing for the caramel stage? Start with an amber colour syrup. With a spoon, drop a few drops into a bowl of ice water, the syrup should instantly harden into a ball.
· Ice water will cool the caramel, causing it to become hard and brittle.

花生芝麻糖不甩
Sticky Rice Nuggets
with Peanuts and Sesame Seeds

一款懷舊小食，材料及製作簡單，毋須特別修飾，每口都給我帶來古早真材實料的滋味。

This delicious snack, simple and without much fanfare, always reminds me of the authentic flavours of the good old days.

材料

（一）

糯米粉	150 克
澄麵	3 平湯匙
糖	3 湯匙
水	約 1/2 杯（125-135 毫升）

（二）

剁碎焗香花生	1/2 杯
炒香芝麻	3 湯匙
糖	3-4 湯匙

做法：

1. 糯米粉、澄麵及糖放大碗內拌勻，加入適量水拌成軟滑粉糰。

2. 預備一隻已掃油的碟，倒入粉糰，撥平。

3. 燒滾半鑊水，隔水蒸熟粉糰約 8-10 分鐘（視乎粉糰厚薄）。

4. 剁碎花生、芝麻及糖盛碗內，拌勻。

5. 把蒸熟糯米糰剪成小件，沾滿花生芝麻成糖不甩。

Note

· 澄麵是從麵粉提煉出來的無筋度麵粉。一般用於製造糕點及米類麵條，加入澄麵的糕點及麵條會較爽口。

· "Wheat starch" is a kind of gluten free flour. It is generally used to make Chinese dim sum and rice noodles. It makes the dough firmer.

Ingredients

(A)
150g glutinous rice flour
3 level tbsp wheat starch
3 tbsp sugar
1/2 cup water, approximately 125-135ml

(B)
1/2 cup chopped roasted peanuts
3 tbsp fried white sesame seeds
3-4 tbsp sugar

Method

1. Place glutinous rice flour, wheat starch and sugar in a mixing bowl. Add water gradually and stir to a soft sticky consistency.

2. Glaze a dish with oil. Spread the sticky rice dough evenly onto the dish.

3. Bring water to a boil in a wok. Steam sticky dough for 8 to 10 minutes until cooked (steam time depends on the thickness of the dough).

4. Mix (B) ingredients together in a bowl.

5. When the sticky dough is cooked, use a pair of scissors, cut the sticky dough into nuggets and coat them evenly with (B) mixture. Serve as snack.

糖合桃
Walnut Kernel Candies

糖合桃、角仔、芋蝦、煎堆等是我每年新年都會自製的應節食品。
當中的糖合桃更會包裝成小禮品送給各親友，物輕情重，希望大家
都有一個豐盛新年。

Each year I follow the tradition of making deep-fried New Year
treats, like crispy dumplings, taro crisps, sesame dumplings, etc.
Walnut Kernel Candies is always a must. I will also pack them as
hearty New Year gifts for friends and relatives.

材料

合桃肉	500 克
冰糖	300 克
水	1 1/2 杯
麥芽糖	1 湯匙
炸油	適量
炒香白芝麻	2 茶匙

做法

1. 冰糖放入水內煮溶，慢火煮片刻至呈現少許黏性，加麥芽糖拌勻。

2. 合桃放 1/4 鑊滾水內煮 1-2 分鐘，取出，濾淨。

3. 把合桃放糖水內浸 6-8 小時，待每粒合核都裹着一層薄薄的糖膠。炸
   前隔去糖水。

4. 用大火燒熱 1/3 鑊油，把合桃放熱油內，然後把火候收慢，浸炸至金
   黃色，慢慢攪動至顏色均勻。取出濾淨油分，撒下白芝麻。

5. 待凍後，才可放密封器皿儲藏。

糖水煮至有少許黏性。
Cook to a light syrup consistency.

Note

· 市面上出售的合桃肉分為有衣或去衣，兩者皆可用。去了衣的合桃肉當然是首選，但必定要新鮮，否則容易變質。

· 而有衣的新鮮合桃肉，經過汆水，也可去除衣的苦澀味。

· Walnut kernels are sold with or without skin. The ones without skin is the better choice, but make sure that they are fresh, as skinless walnut kernels can go bad easily.

· Blanching the fresh walnut kernels with skin can minimize the bitter taste of the skin.

Ingredients

500g walnut kernels
300g rock sugar
1 1/2 cups water
1 tbsp maltose
oil for deep-frying
2 tsp fried white sesame seeds

Method

1. Bring water and sugar to a boil until sugar is dissolved. Lower heat and boil a little longer until it has a light syrup consistency. Stir in maltose.

2. Blanch walnut kernels in 1/4 wok boiling water for 1-2 minutes. Remove and drain well.

3. Add walnut kernels to the light syrup and set aside for 6 to 8 hours until the walnut kernels are coated with a thin film of syrup. Strain just before deep-frying.

4. Heat 1/3 wok oil, lower walnut kernels into hot oil. Reduce to medium-low heat and deep-fry until golden and crispy, stirring gently from time to time to obtain even colouring. Remove walnut kernels and drain well. Sprinkle with sesame seeds.

5. Leave to cool completely before storing in air-tight containers.

反沙芋條
Crystallized Taro Strips

秋冬是芋頭當造的季節，很容易便可買到既香又鬆化的芋頭，這些良品就是用來炮製反沙芋條的最必要元素。反沙的做法是先把糖煮溶成糖漿，加入預先走油至熟透的芋條，慢慢兜炒至每條芋條都裹上了糖漿，並轉化成為結晶體狀。反沙做得好，吃時外脆內鬆化，隨時可令人食上三、四條，成為潮式晚宴的完美結局。

When in season during fall and winter, taro is best for making Crystallized Taro Strips, because of its special fragrance and soft starchy texture. To obtain the crystallized coating, cook the sugar until syrupy. Coat fried taro strips evenly with the syrup, continue tossing until white sugar crystals appear and are coating the taro strips. When prepared properly, the taro strip is crispy on the outside and has a soft and starchy centre. It is simply irresistible! One can easily eat three to four strips in one sitting. This dessert rounds up a scrumptious Chaozhou dinner.

材料

芋頭	300 克
白砂糖	120 克
水	1/8 杯
乾葱	1 粒（略拍）

做法

1. 芋頭去皮，沖淨，切 1.5 厘米 x 1.5 厘米 x 6 厘米長條。

2. 燒 1/4 鑊油，油微熱時放下芋條，用中慢火炸熟，及至外層乾身，取出。

3. 鑊內剩 1 湯匙油炒香乾葱，然後棄掉乾葱。加水及糖煮滾，拌溶糖後繼續煮至透明糖漿狀，不可以過稠。

4. 把炸好的芋條放糖漿內，不停與糖漿兜炒均勻，繼續炒至裹着芋條外層的糖變成白色「沙狀」，趁熱品嘗。

Ingredients

300g taro
120g sugar
1/8 cup water
1 shallot, lightly crushed

Method

1. Peel taro, rinse and cut into 1.5cm x 1.5cm x 6cm long strips.

2. Heat 1/4 wok oil, deep-fry taro strips in medium heat until cooked and slightly dry on the outside. Remove and drain.

3. Drain away oil in the wok, leaving 1 tbsp behind. Sauté shallot until fragrant, then discard. Add water and sugar and bring to a boil. Stir until it becomes light syrup. Do not overcook.

4. Add fried taro strips to the syrup, mixing well. Toss until the syrup coating on the taro strips becomes crystallized. Serve at once.

醋
Vinegar

想要做出酸甜、酸辣菜式，又或炮製醃漬菜，又怎能沒有醋的幫忙？

食用醋的種類繁多，全中國各地各省都有不同的品種，你又可知當中分別或如何使用？

How can you do without vinegar when cooking sweet and sour dishes, hot and sour dishes or pickled vegetables?

There are many different varieties of vinegar throughout China, which ones are more suitable for your cooking needs?

大閘蟹醋
Vinegar Dip for Crab

材料

鎮江香醋	1/2 杯
薑米	1/4 杯
黃糖	2-3 湯匙
生抽	1/2-1 湯匙

做法

1. 鎮江香醋與薑米同放小鍋內,用中慢火煮至薑味溢出。

2. 拌入適量黃糖,讓醋有些甜味。

3. 最後加入適量生抽,調校至甜酸帶微鹹、芳香濃稠的蟹醋。

Ingredients

1/2 cup Zhenjiang dark vinegar

1/4 cup ginger rice

2-3 tbsp brown sugar

1/2-1 tbsp light soy sauce

Method

1. Place dark vinegar and ginger rice in a small saucepan, slowly bring to a boil, simmer until ginger flavour has infused into the vinegar.

2. Stir in enough brown sugar for a mild sweet taste.

3. Add enough light soy sauce to blend. This vinegar dip has a blend of vinegary, sweet and savoury taste, as well as a rich ginger aroma.

Note

· 薑米是把薑切成很幼細似米形狀,沒有汁液瀉出,仍然保存薑的辛辣味。若用磨碎的薑茸,味道與口感都不一樣。

· What is ginger rice? It is in fact ginger finely chopped to the size of rice. These little particles of ginger retain their flavour and juice completely. If grated ginger is used instead, both flavour and texture of this dip will not be the same.

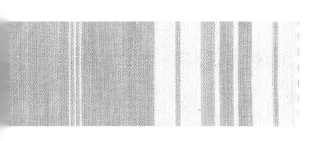

蒜茸米醋
Minced Garlic Vinegar Dip

材料

白米醋	3 湯匙
蒜茸	2 茶匙
紅辣椒碎	1 茶匙
鹽及糖	少許

做法

白米醋加入少許鹽及糖拌勻，再加入蒜茸及紅椒碎即成。

Ingredients

3 tbsp rice vinegar
2 tsp finely chopped garlic
1 tsp finely chopped red chilli
salt and sugar to taste

Method

Mix rice vinegar with a little salt and sugar to taste. Stir in chopped garlic and chilli.

Note

· 用新鮮手剁的蒜茸比用機器磨的或樽庄的蒜茸更清香。
· 如果白米醋的酸味過濃，可用少許凍飲用水稀釋。
· 白米醋加入了少許鹽及糖，可提升整個蒜茸米醋的味道。

· Freshly chopped garlic has a more refreshing and pungent flavor than those sold in jars or minced in a blender.
· If desired, you can lighten the acidity of rice vinegar by adding a little cooled boiled water.
· Adding a little salt and sugar to the vinegar can boost the flavour of the dip.

餃子醋
Vinegar Dip for Dumplings

材料

鎮江香醋　　4 湯匙
幼薑絲　　　2 湯匙

做法

把新鮮切好的薑絲放小碟內，加入適量鎮江香醋
拌一拌。隨即可蘸餃子。

Ingredients

4 tbsp Zhenjiang dark vinegar
2 tbsp finely shredded ginger

Method

Place freshly shredded ginger in a small dish.
Add dark vinegar and mix together just before
serving.

Note

· 新鮮拌起的餃子醋特別清香，融
　入了辛香的薑味；酸咪咪的醋配
　素或肉餡餃子，可有助消化。

· Freshly prepared Vinegar
　Dip for Dumplings has a
　refreshing ginger flavour. The
　vinegary taste complements
　both vegetable and meat
　dumplings, and eases
　digestion at the same time.

松鼠魚
Squirrel Fish in Sweet and Sour Sauce

明明是魚，為何會被冠以「松鼠」的名稱呢？其實這是用來形容魚因為切割方法令它在煮熟後翹起，形態就彷如松鼠尾巴一樣，因此而命名。

What is "Squirrel Fish"? It is in fact a fish, cut and fried to shape like a squirrel with a bushy tail.

材料

桂花魚	1 條（500-550 克）
蛋	1 個（拌勻）
生粉或麵粉	1 杯
焗香松子仁	2 湯匙

Ingredients

- 1 fresh mandarin fish (500-550g)
- 1 beaten egg
- 1 cup potato starch or plain flour
- 2 tbsp roasted pine nuts

醃料

鹽	3/4 茶匙
胡椒粉	少許

Marinade

- 3/4 tsp salt
- a little pepper

芡汁

（一）

青椒及紅椒	2 湯匙（切幼粒）
菠蘿	1 片（切粒）
蒜茸	1 茶匙

（二）

水	3/4 杯
白醋	3 湯匙
茄汁	3 湯匙
鹽	1/4 茶匙
糖	1 1/2-2 湯匙
生粉	2 茶匙
橙紅粉	適量（隨意）

Sauce

(A)

- 2 tbsp red and green pepper, diced
- 1 pineapple ring, diced
- 1 tsp chopped garlic

(B)

- 3/4 cup water
- 3 tbsp white vinegar
- 3 tbsp tomato ketchup
- 1/4 tsp salt
- 1 1/2-2 tbsp sugar
- 2 tsp potato starch
- a little orange-red colouring, optional

做法

1. 桂花魚去鱗、去鰓，用刀從魚之背部片成雙飛，切去魚骨，沖淨，抹乾，於魚肉上剧欖角形紋，搽上鹽及胡椒粉。將魚沾蛋液，再均勻撲上乾粉。

2. 燒半鑊油，把魚捏成松鼠形放入熱油內炸熟。將魚取出，再燒熱油，將桂花魚回鑊多炸一次至香脆，取出盛碟。

3. 燒 1 湯匙油，爆香芡汁（一），倒下芡汁（二）煮成甜酸汁。

4. 將汁料淋於松鼠形的桂花魚上，撒下松子仁，趁熱享用。

焗松子仁

把松子仁放焗盤內，用中慢火焗爐（攝氏 180度）焗至金黃香脆，不時拌一拌使顏色均勻。

Method

1. Scale and clean fish, slit open along the back and remove centre bones, rinse and wipe dry. Cut criss-cross pattern on the meat, rub fish with marinade. Dip whole fish in beaten egg, then coat evenly with starch or flour.

2. Heat 1/2 wok oil to medium-hot. Shape fish like a squirrel, lower fish into oil and deep-fry until golden in colour. When fish is cooked, remove and drain. Heat oil again and fry fish a second time until crispy. Remove fish, drain well and arrange on a plate.

3. Heat 1 tbsp oil, stir-fry sauce (A) until fragrant, add sauce (B) and bring to a boil.

4. Pour sauce over fried whole fish, sprinkle with roasted pine nuts and serve.

Roast pine nuts

Place pine nuts on a roasting tray and roast in a medium-hot oven (180°C), turning them once in a while until evenly golden in colour.

Note

· 片魚肉的刀一定要鋒利，否則很難切到齊整的魚片。

· 要將魚柳上多餘的粉拍出，炸後的魚肉格仔紋才突出，也可避免炸油渾濁。

· 放魚柳下鍋時，確保要用手執着魚柳的首尾兩端以固定形態，讓魚肉炸起後形態神似松鼠。

· 怎樣儲存炸油和保鮮？把炸油翻熱蒸發水分和氣味。鐵篩內鋪廚房萬用紙或蒸漏咖啡紙，倒入炸油濾淨雜物，待涼後儲起，用於日常烹煮，盡快用完。

· Use a sharp knife to slice the fish to obtain a uniform cut.

· Shake off the excess flour from the fish before deep-frying so the criss-cross pattern appears. There will also be less sediment left in the oil.

· Hold the fish at both ends as you lower it into the hot oil, so it will retain its squirrel-like shape when cooked.

· Heat up the used oil to vaporize its moisture and smell. Strain used oil through paper towel or coffee filtered paper in a metal sieve. Leave to cool and store in a container. Use up this oil for daily frying as soon as possible.

辣泡菜
Spicy Pickled Vegetables

這道爽脆泡菜味道鹹、酸,不甜,再加上四川花椒,令到味道帶點輕微的麻香。如用作餐前小吃,惹味醒胃,或是用來伴飯,當心吃完一碗又一碗!

These crunchy pickles contain an assortment of flavours.....salty and sour. You will also experience a tingling sensation from the spicy Sichuan peppercorns. They truly wake up your appetite before a full meal.

材料

椰菜	300 克
白蘿蔔	300 克
尖紅辣椒	2-3 隻
蒜頭	4 粒(剁碎)
薑	4 片(剁碎)
乾辣椒	2-3 隻

泡菜滷汁

米醋	1/4 杯
鹽	1 1/2 湯匙
糖	1 湯匙
凍飲用水	1 1/2 杯
炒香紅川椒粒	2-3 茶匙

做法

1. 白蘿蔔去皮與椰菜沖淨抹乾,切成小塊。

2. 尖紅辣椒切段,蒜頭略拍,薑切片;乾辣椒沖淨後剪去頭部,切段。

3. 泡菜滷汁拌勻至鹽及糖溶化。

4. 全部材料放入泡菜滷汁內拌勻,蓋好放雪櫃最少兩天。

Note

· 首選玻璃、瓦質或瓷器器皿盛泡菜；其次是不鏽鋼，忌用鐵、鋁或其他金屬，以免起化學作用。
· 用滾水沖淨器皿，晾乾才可用。
· 選擇新鮮當造的蔬果做泡菜，例如夏令瓜果、秋冬根莖蔬菜。

· It is best to use glass, clay or chinaware containers to store pickles. Stainless steel can be an alternative. Do not use aluminum or other kinds of metal as it might cause chemical reaction.
· Rinse container with boiling water, drain until completely dry before use.
· Always use the freshest vegetables and fruits in season for pickling, like squashes and fruits in summer, vegetables and root vegetables in winter.

Ingredients

300g cabbage
300g turnip
2-3 hot red chillies
4 cloves garlic, chopped
4 slices ginger, chopped
2-3 dried red chillies

Pickling solution

1/4 cup rice vinegar
1 1/2 tbsp salt
1 tbsp sugar
1 1/2 cups cooled boiled water
2-3 tsp fried Sichuan red peppercorns

Method

1. Peel turnip, rinse and wipe dry turnip and cabbage, cut into smaller pieces.

2. Cut red chillies into sections, lightly crush garlic, slice ginger; rinse dry red chillies, cut off stalks and caps, then cut into sections.

3. Mix pickling solution well until salt and sugar are completely dissolved.

4. Add all vegetables into pickling solution, mix well. Cover and set aside in the fridge for at least 2 days before serving.

越式雞肉法包
Baguette with Chicken Julienne

我第一次遇上「越式法包」並不是在越南或法國，而是溫哥華。長居彼邦的姐姐深知我這個小妹饞嘴，因此總會為我多加留意有特色的食品。有次又到訪姐姐，她特意驅車前往烈治文，極力推薦一間小店的「越式法包」給我。進店後我們分別點了雞肉和豬肉餡的法包三文治。不消一會三文治送到，麵包烘得外脆內軟，餡料都細心的撕成幼絲，加上大量的泡菜絲及芫茜，滿滿的塞在麵包內，必需要盡量張大嘴巴才可把麵包連同餡料一併放入口內。與老闆娘攀談起來，原來她是我的忠心「粉絲」，大家一見如故，大談食經。老闆娘還提點要令這個越式法包更美味，一定要加點兒指天椒碎。向來不嗜辣的我起初有點抗拒，但嘗過後，發覺指天椒真的能把本來已很美味的三文治味道更提昇！

My first encounter with this "Vietnamese Baguette" was neither in Vietnam nor in France. It was introduced to me by my elder sister in Vancouver. I tried this baguette for the first time in a small restaurant in Richmond. We ordered one with chicken and one with pork. Within minutes, two baguettes piled high with shredded meat, pickles and coriander arrived at our table. They were so big that it was not easy biting into them. I also found out that the owner was one of my cooking fans. So we spent a lot of time talking about food and cooking. She suggested adding a little bird's eye chilli to heighten the flavour of this baguette. Not a big fan of bird's eye chilli, I was reluctant to try that at first. But she was right, a touch of heat really lifted the whole flavour!

材料

雞胸肉	300 克
洋葱	1/4 個（切幼條）
西芹	1/2 枝（切段，略拍）
魚露	2 湯匙
糖	1 茶匙
黑胡椒碎	1 茶匙
長法式麵包	1 條（切 4 段）
豬肝醬或雞肝醬	適量
沙律醬	適量
芫茜	1 束（清洗，去根部）
紅辣椒或指天椒	1-2 隻（切薄片）

泡菜絲

細紅蘿蔔	1/2 條（刨絲）
白蘿蔔	1/4 條（刨絲）
白醋	2/3 杯
白砂糖	1/2 杯
鹽	1/4 茶匙

做法

1. 泡菜絲：白醋、糖及鹽拌勻，分為兩份。紅及白蘿蔔絲分別放糖醋內，放雪櫃醃泡 30 分鐘，用時取出隔淨汁液。

2. 洋葱、西芹、魚露、糖及黑胡椒碎拌勻，搽勻雞胸肉醃 1 小時。雞胸肉放中火烤爐烤至兩面甘香及肉熟（約 10-15 分鐘）。取出略凍後，撕成幼條狀。

3. 法式麵包用麵包刀橫剝開，放焗爐或烤爐烘脆。

4. 烘脆法式麵包底邊搽上肝醬，加入泡菜絲、雞肉條、沙律醬、芫茜及紅辣椒（隨意），蓋上另一邊法式麵包，把餡料夾於中間。享用時配香濃越式滴滴咖啡，特顯風味。

Ingredients

300g boneless chicken breast
1/4 small onion, cut into strips
1/2 stalk celery, sectioned and lightly crushed
2 tbsp fish sauce
1 tsp sugar
1 tsp crushed black peppercorns
1 long baguette, cut into 4 sections
a few tbsp pork or chicken pâté
a few tbsp salad dressing
1 small bunch coriander, washed and roots removed
1-2 red chillies or bird's eye chillies, thinly sliced

Pickles

1/2 small carrot, grated
1/4 white turnip, grated
2/3 cup white vinegar
1/2 cup sugar
1/4 tsp salt

Method

1. Pickles: Mix white vinegar, sugar and salt together until completely dissolved. Divide mixture into 2 mixing bowls. Add grated carrot and turnip separately to the 2 bowls and set in fridge for 30 minutes. Strain well before use.

2. Mix onion and celery with fish sauce, sugar and black peppercorns. Rub marinade into the chicken breast and set aside for 1 hour. Grill chicken breast in a medium hot grill until meat is cooked and slightly charred (approximately 10-15 minutes). Remove, cool and tear into julienne.

3. Using a bread knife, cut the baguette horizontally and toast lightly.

4. Spread pâté on bottom side of baguette, then add vegetables pickles, chicken julienne, salad dressing, coriander and chilli (optional). Put on the baguette top. Serve with Vietnamese coffee.

越式滴滴咖啡

Vietnamese "Dripping" Coffee

越南咖啡的沖泡方法是把咖啡粉放在滴餾咖啡壺內，杯內加入適量煉奶，然後把滴餾咖啡壺放置杯上，倒入沸水，讓咖啡一滴一滴的流進杯內。整個過程緩慢進行，不由催促，否則咖啡便不夠濃了。故此，越式滴滴咖啡總是給我閒適的感覺，邊等邊望街發呆，讓腦筋靜下來，放鬆心情，享受悠閒一刻。

Vietnamese coffee is filtered coffee made in a small traditional coffee maker. Condensed milk is first added to the cup. A strainer with ground coffee is placed on top. Hot water is added and left to drip through, slowly brewing the coffee to a full aroma. That's no rush! Relax and enjoy the leisure moment while the coffee is brewing.

材料

越南咖啡粉
熱水
煉奶
一套越式滴餾咖啡壺
咖啡杯

做法

1. 把適量煉奶放咖啡杯內。

2. 滴餾壺放咖啡杯上，加入咖啡粉，輕輕撥平，上面放篩子。先加入少許熱水泡一泡咖啡粉（約30秒），然後把篩子輕輕向下壓，不可太實。再加熱水。蓋上蓋子，待咖啡遂少滴餾入杯內。

3. 享用時把煉奶與咖啡拌勻。

Ingredients

ground Vietnamese coffee
hot water
condensed milk
a set of Vietnamese coffee maker
coffee cup

Method

1. Put condensed milk in the coffee cup.

2. Put the dripping funnel over the coffee cup. Add enough ground Vietnamese coffee (according to individual taste) to the dripping funnel, spread out evenly. Place the small metal plate on top. Add a little hot water to the ground coffee and wait for 30 seconds. Lightly weigh down the ground coffee with the small metal plate. Add hot water to the ground coffee, cover with the lid and leave the coffee to drip through.

3. Stir well with the condensed milk to serve.

糖醋排骨
Sweet and Sour Spareribs

甜酸味的菜餚在中國不同省份派系的菜式中都有，但當中的滋味又總有點不同。其中一個原因就是使用了不同品種的醋，令其甜酸味各有特色、各有千秋。像這道「糖醋排骨」是上海菜中的冷盤，使用了鎮江黑醋，醋味較濃厚，十分醒胃。

There are many versions of sweet and sour sauce throughout the different regions of China. Each has its own unique taste due to the different kind of vinegar used. This "Sweet and Sour Spareribs" Shanghai style is mostly served as starter. It arouses the appetite with the rich and pungent taste of Zhenjiang dark vinegar.

材料

腩排	500 克（斬成 3 厘米小段）
薑	4 片
葱	1 條
黃酒	3 湯匙
水	約 300 毫升（蓋過小排骨）
糖	2-3 湯匙
生抽	1 1/2 湯匙
鎮江黑醋	4-5 湯匙
八角	2 粒

做法

1. 小排骨沖淨及抹乾；燒 1 湯匙油，爆香薑及葱，放入小排骨兜炒至微黃色，灒下黃酒，加適量水蓋過小排骨，加入其餘材料，先煮滾，改用慢火加蓋燜腍（約 40-45 分鐘）。期間要把小排骨略兜炒避免黏鍋底，需要時加適量水。

2. 小排骨燜腍後，改調中猛火快手兜炒至汁濃稠。

Ingredients

500g spareribs, chopped into 3cm sections

4 slices ginger

1 stalk spring onion

3 tbsp brown wine

approximately 300ml water (enough to cover the spareribs)

2-3 tbsp sugar

1 1/2 tbsp light soy sauce

4-5 tbsp Zhenjiang dark vinegar

2 star anise

Method

1. Rinse and wipe dry spareribs. Heat 1 tbsp oil, sauté ginger and spring onion until fragrant, add spareribs and fry until light brown. Add brown wine and mix well. Add enough water to cover the spareribs, then add the rest of the seasonings. Bring back to a boil, lower heat to simmer, cover and cook until spareribs are tender (approximately 40-45 minutes). From time to time, stir the spareribs to prevent sticking. Add more water if necessary.

2. When spareribs are cooked, toss over high heat until well glazed.

Note

· 最後的步驟不可馬虎，要把糖醋汁兜炒至晶亮糖膠狀裹着每一顆小排骨。

· Last but not least, the sweet and sour sauce must be reduced to a syrupy state glazing each piece of sparerib.

京都肉排
Kingtao Pork Chop

京都肉排之所以成為廣受歡迎的名菜，除了其討好的甜酸味外，令它與一般甜酸菜不同之處，就是加入了略帶煙燻味的噫汁，令味道更添層次。記得有一次我利用這個京都汁煮了一味京都蝦給一位小朋友吃，想不到事隔多年，該名小朋友已長大成人，並成了一名醫生，但至今他仍銘記這道菜，而我就是「京都蝦 Annie 姐姐」了！

Kingtao Pork Chop is an all-time favourite dish. The smoky taste from the Worcestershire sauce gives it a unique flavour unlike any other sweet and sour sauces. I once cooked prawns with this sauce for a young friend of mine. Now a prominent doctor, he still remembers "Sister Annie's Kingtao Prawns".

材料

無骨豬扒或枚頭肉	400 克
生粉	1/2 湯匙
乾葱	1 粒（切碎）

醃料

（一）

梳打食粉	1/4 茶匙
水	1 湯匙

（二）

五香粉	1/4 茶匙
鹽	1/2 茶匙
糖	1/2 茶匙
生粉	1 茶匙
生抽	1 茶匙
玫瑰露酒	1 茶匙
蛋液	1 湯匙
乾葱茸	1 茶匙

芡汁

鹽	1/8 茶匙
糖	1 1/2-2 湯匙
水	3 湯匙
大紅浙醋	3 湯匙
茄汁	1 湯匙
噫汁	1 茶匙

做法

1. 用刀背拍鬆豬扒，再切小件。先與醃料（一）拌勻，再加醃料（二）拌勻，待 1 小時。

2. 燒 1/4 鑊油，豬扒濾淨醃料，撒上生粉，放熱油內炸熟，取出濾淨油分。

3. 燒 1 湯匙油，爆香乾葱茸，加芡汁煮至濃香，把炸熟之豬扒回鑊，與芡汁兜勻至呈糖膠狀。

Ingredients

400g boneless pork chop or pork loin
1/2 tbsp potato starch
1 shallot, chopped

Marinade

(A)
1/4 tsp cooking soda
1 tbsp water

(B)
1/4 tsp five spiced powder
1/2 tsp salt
1/2 tsp sugar
1 tsp potato starch
1 tsp light soy sauce
1 tsp Mei Kuei Lu Chiew
1 tbsp beaten egg
1 tsp chopped shallot

Sauce

1/8 tsp salt
1 1/2-2 tbsp sugar
3 tbsp water
3 tbsp red vinegar
1 tbsp ketchup
1 tsp Worcestershire sauce

Method

1. Pound pork chop with the back of the chopper, then cut into bite-size pieces. Mix with marinade (A) until well absorbed, then mix with (B). Set aside for 1 hour.

2. Heat 1/4 wok oil, drain off marinade and lightly dust pork pieces with potato starch, deep-fry in medium-hot oil until cooked. Remove and drain.

3. Heat 1 tbsp oil in the wok, sauté chopped shallot, add sauce and cook until a little syrupy, then add pork pieces and toss until well glazed.

酒
Wine

酒可提鮮、增香、添味。在烹煮時，酒精遇熱產生片刻高溫並散發出誘人香氣，令人食欲大增。這就是料酒的重要功勞之一了！

Wine plays a vital role in enhancing the flavour of many dishes. When combined with other ingredients during cooking, wine emits a unique aroma which invigorates one's appetite. That is the magic of cooking with wine!

自家釀青梅酒
Homemade Green Plum Wine

自家釀酒的趣味在於不同配搭，可應個人口味調校。
你可嘗試使用不同味道和酒精含量的酒、不同生熟程度
的青梅，與及使用不同種類的糖，如冰糖、有機黃砂糖
等等。當然酒、糖及青梅的配方比例會影響梅酒的味道。
自釀梅酒是一件很好玩的事！快來釀製有你個人風格的
梅酒吧！

The joy in brewing your own wine lies in the freedom to
mix and match ingredients to your own liking.
You can mix wines of different flavour and alcohol
content with sour or ripe green plums, and with different
kinds of sugar, such as rock sugar, organic brown
sugar, etc. The reward comes at the end of the day. The
character of the wine bears your signature and your own
customized formula!

材料

花雕酒	500 毫升
新鮮青梅	400 克
冰糖	300 克

用具

潔淨玻璃瓶（把玻璃瓶放入滾水內焓 15 分鐘，瀝乾備用；或 將洗
淨瀝乾後的玻璃瓶注入少許酒盪勻，然後倒掉）

Ingredients

500ml Hua Diao wine
400g green plums
300g rock sugar

Tools

clean glass jar
Boil glass jar for 15 minutes, drain well before use.
Or
Wash glass jar and drain well, add some wine, swirl round all sides of
the jar, pour away wine and drain well before use.

做法

1. 青梅沖淨，除去蒂部。抹乾或吹乾青梅至沒有水分，用鐵針於青梅肉上插幾下，可助青梅味滲出。

2. 將冰糖與青梅分層放入玻璃瓶內，注入花雕酒蓋過面，蓋上玻璃瓶蓋，封實。

3. 放在陰涼位置儲存，待六個月後便成醇香清甜的自家釀青梅酒。

Method

1. Rinse green plums, remove the stalk. Wipe dry or leave to air-dry. Make sure there is no more moisture left on the plums. Pierce plums several times to let the flavour out during brewing.

2. Add alternate layers of rock sugar and green plums into the glass jar. Add Hua Diao wine to cover completely.

3. Store the glass jar in a cool place, away from direct sunlight. In 6 months, the plum wine will be ready for drinking.

青梅酒果凍
Green Plum Wine Jello

青梅酒最適宜加冰凍飲；梅味香、酒味醇，每啖都
清涼透心。把它製成果凍，更多添 QQ 的口感！
Green plum wine is best served chilled. It offers you
the refreshing fragrance of green plums, as well as
the silkiness of the wine. Try making it into fruit jello,
the taste is just amazing!

材料

青梅酒	1 1/2-1 3/4 杯
魚膠粉	2 平湯匙（20 克）
水	1/2 杯
浸過酒的青梅	數粒

做法

1. 把 2 平湯匙魚膠粉舀於 1/4 杯清水內，待片刻至魚膠粉吸入水分而膨脹。

2. 把其餘 1/4 杯水煮滾，放入已膨脹的魚膠粉拌至溶化，待至略凍，與青梅酒拌勻。

3. 每隻小杯內放一粒浸過酒的青梅，倒入以上梅酒，放入雪櫃內凝固成果凍。

Ingredients

1 1/2-1 3/4 cups homemade plum wine
2 level tbsp gelatine powder, 20g
1/2 cup water
a few green plums from the plum wine

Method

1. Sprinkle gelatine powder over 1/4 cup water, set aside until the water is well absorbed.

2. Bring the other 1/4 cup water to a boil. Add gelatine mixture and stir until completely dissolved. Let cool to room temperature, then mix with plum wine.

3. Place a green plum into each serving glass. Fill up with plum wine gelatine. Chill and set in fridge.

Note

· 用 1 湯匙魚膠粉（10 克）對 1 杯液體可做成一般質感的果凍。若喜歡軟滑口感的果凍，可多加 3 湯匙液體。

· Use 1 level tbsp (10g) gelatine powder to 1 cup liquid for normal textured jelly. If desired, add an extra 3 tbsp liquid for a softer jelly.

話梅花雕蝦
Sour Plum Hau Diao Prawns

花雕酒味道香醇與海鮮十分合配。把灼熟的蝦浸於剛泡溫暖的花雕酒內，端上檯時，酒香與及蝦的鮮味迎面而來，芳香撲鼻！

Hua Diao wine complements most seafood. Try immersing cooked prawns in warm Hua Diao wine for a few minutes before serving. The wine will permeate the prawns, accentuating both the taste and the aroma!

材料

（一）

活中蝦	300 克
花雕酒	1/4 杯

（二）

靚話梅	6-8 粒
花雕酒	1 杯

做法

1. 話梅放入花雕酒內浸泡 10 分鐘。

2. 中蝦沖淨，隔乾備用。烹調前 5 分鐘拌入 1/4 杯花雕酒浸蝦。

3. 將話梅花雕酒慢火煮熱，試味。

4. 燒大半鑊水，放下中蝦灼熟，取出隔淨水分。

5. 中蝦盛於深碟內，注入燙熱之話梅花雕酒，即可享用。

Ingredients

(A)
300g medium size live prawns
1/4 cup Hua Diao wine

(B)
6-8 preserved sour plums
1 cup Hua Diao wine

Method

1. Combine preserved sour plums with 1 cup Hua Diao wine. Leave to steep for 10 minutes.

2. Rinse live prawns, drain well. Add 1/4 cup Hua Diao wine to the prawns 5 minutes before cooking. Leave to marinate.

3. Heat up the steeped Hua Diao wine in low heat until hot.

4. Bring half wok of water to a boil, add in prawns. When the water comes back to a boil, the prawns are now cooked. Remove and drain well. Be careful not to overcook prawns.

5. Place prawns on a plate, pour heated Hua Diao wine over and serve.

Note

· 蝦要先灼熟，瀝乾水分，然後放入話梅花雕酒內，而不是把鮮蝦放入話梅花雕酒內煮熟，否則會使菜式味道偏腥。

· 此菜式蝦味鮮而酒味香甜，若要酒味濃些，可隨意減少話梅。

· Cook prawns separately, drain well before adding to the sour plum flavoured wine. Do not cook prawns in the wine as this will make the wine taste fishy.

· This dish offers you the "sweet freshness" of both prawn and wine. Use one or two less preserved sour plums if a richer wine flavour is preferred.

醉轉彎
Drunken Chicken Wings

剛認識「醉轉彎」時，光看名字都未有猜透是甚麼樣的一道菜，上桌後便知道原來是雞翅膀的轉彎位置。這個部位肉不多，但懂得欣賞的人就是喜歡它的嫩滑。特別是用來做冷菜，翅膀吸收了花雕的酒香，便更顯香滑！

Chicken wings have a delicate texture which can absorb the fragrance of the wine easily. Drunken Chicken Wings will make an excellent cold starter.

材料

冰鮮雞翼	500 克
薑汁	1 湯匙
鹽	2 茶匙

花雕醉雞汁

蒸雞汁	1/2-2/3 杯
花雕酒	1/2-2/3 杯

做法

1. 雞翼沖淨，放熱水內拖水 3 至 5 分鐘，取出，用清水沖淨，去除油脂。

2. 雞翼抹乾後，先用薑汁搽勻，再搽鹽，醃 15 分鐘。

3. 把雞翼排放碟上，隔水用中火蒸 8 分鐘。不用打開鍋蓋，待 10 分鐘，使雞翼熟透。取出蒸好的雞翼，用飲用冰水浸涼，使皮爽脆；蒸雞汁留起。

4. 花雕醉雞汁拌勻，試味，可隨意加多些花雕酒使酒味更濃。

5. 把雞翼浸於花雕醉雞汁內，翌日成涼菜品嘗。

Ingredients

500g chilled chicken wings
1 tbsp ginger juice
2 tsp salt

Wine stock

1/2-2/3 cup stock from steaming the chicken wings
1/2-2/3 cup Hua Diao wine

Method

1. Rinse chicken wings. Blanch for 3-5 minutes, take out and rinse well to remove surface grease.

2. Wipe dry chicken wings. Rub with ginger juice, then with salt and leave to marinate for 15 minutes.

3. Arrange chicken wings on a plate, cover and steam over medium heat for 8 minutes. Leave the cover on and wait for another 10 minutes, so that the chicken wings are completely cooked. Remove chicken wings and plunge into ice-cool boiled water until completely cool, this will crisp up the skin. Save the chicken stock.

4. Blend Hua Diao wine with the chicken stock, adjust taste. If desired, add more Hua Diao wine for a richer wine flavour.

5. Return chicken wings to the wine stock and leave in the fridge overnight. Serve cold as starter the next day.

Note

· 冰鮮雞翼價錢較新鮮的便宜，如處理得宜，味道不比新鮮的遜色。我用花雕酒和蒸雞汁將雞翼浸至入味，味道一級棒！
· 不要用猛火蒸雞翼或蒸過時，否則雞皮會起皺紋及近骹部位爆裂。

· Chilled chicken wings are less expensive than the fresh ones. When properly treated, their flavour is just as good. I immerse the chicken wings in a mixture of Hua Diao wine and the concentrated stock from the steamed chicken wings, the flavour is delightful!
· Do not steam the chicken wings over high heat or over steaming them. This will cause the skin to wrinkle and split at the joints.

嫲嫲月婆雞酒
Chicken in Brown Wine Broth

「嫲嫲月婆雞酒」是我嫲嫲的拿手好菜。這滋養補品不限於「月婆」專利。嫲嫲知道我很喜歡吃，每逢於一家團聚的晚飯及我的生日，這味雞酒必定不可少。This wholesome dish, traditionally prepared for women who have just given birth, is also good for everybody. It is not only my grandma's signature dish, but is also my favourite. She always cooked it for family gatherings, as well as my birthdays.

材料

光雞	1 隻（1.2 千克）
薑	150-200 克
木耳	2-3 朵（20 克）
黃酒	2 杯
水	5-6 杯
鹽	少許
雞蛋 (每人 1 隻)	

做法

1. 光雞洗淨及抹乾，斬件備用。

2. 薑去皮，沖淨，切片。

3. 木耳沖淨，用水浸透，取出剪去硬端，然後撕碎。飛水，再沖淨。

4. 鍋內燒熱 2 湯匙油，先爆香薑片，加入雞件爆炒至微金黃，下木耳，兜炒數下，注入黃酒邊煮邊兜炒至均勻。倒入水再煮滾，改用中慢火，蓋好煮 25 至 30 分鐘，加適量鹽調味。

5. 雞蛋用油煎成荷包蛋。

6. 食時把荷包蛋放進月婆雞酒內煮滾，一起品嘗，以增添營養及味道。

Ingredients

1 chicken (1.2kg)
150-200g ginger
2-3 wood ear fungus (20g)
2 cups brown wine
5-6 cups water
a little salt to season
eggs (one for each person)

Method

1. Rinse and wipe dry chicken, chop into bite-size pieces.

2. Peel ginger, rinse and slice.

3. Rinse and soak wood ear fungus until soft. Trim off hard ends. Tear into small pieces. Blanch and rinse well.

4. In a pot, heat 2 tbsp oil, sauté ginger until fragrant, add chicken pieces and fry until light golden colour. Add wood ear fungus, toss for a while. Add brown wine and fry until well mixed. Add water and bring to a boil. Lower heat and simmer for 25 to 30 minutes. Season to taste with a little salt.

5. Fry eggs, one for each person.

6. To serve: Add fried eggs to chicken and wine broth. Heat up and serve hot. Eggs add more nutrients and flavour to the dish.

Note

· 其他家鄉的做法可加入豬肉、豬肝、豬粉腸、紅棗或冬菇等等。

· 各處鄉村各處例，有些食譜會用糯米酒，但我喜歡黃酒的香氣，你不妨試試。

· Other ingredients like pork, pork liver, pork intestine, red dates, Chinese mushrooms can be added to the broth, according to different traditions.

· Different kinds of wine can be used according to one's liking and tradition. And I prefer brown wine for its unique aroma.

Annie Wong
黃婉瑩

烹飪導師
二十多年教學經驗，一顆熾熱之心仍然不倦，喜愛與人分享烹調心得，學生來自世界各地。

著名電視烹飪主持
擔任電視烹飪節目主持愈十多年，粉絲群不分年齡界別，無遠弗界至全球能收看華語節目的地方，極受歡迎。

食物全接觸
1980年開始為飲食雜誌、廣告擔任食物造型師，並擔任眾多著名食物品牌及廚具的飲食顧問

被邀請到多個國家作中菜示範，推廣中國的烹飪藝術

撰寫食譜及任電台嘉賓主持

全心全意愛烹飪
飲食無國界，Annie愛到世界各地尋找地道菜式及特殊食材，凡西菜及東南亞菜式亦是她的拿手好戲。

Cookery Instructor
With more than 20 years of experience as a cookery instructor, Annie is passionate about sharing cooking tips with others. She has students all over the world.

Famous Cookery Presenter
Annie has been hosting popular TV cookery programs for more than ten years. She has fans in every corner of the world which broadcasts Chinese television.

Exploring every aspect of food
Since 1980, Annie has been a food stylist for food magazines, food ads and TV commercials, as well as food consultant to famous food products and cooking equipment.

Invited to demonstrate Chinese cuisines in many countries, Annie is a true ambassador for the art of Chinese cooking.

Infatuation with Food
Always fascinated with savouring unique dishes and ingredients from different parts of the world, Annie has become an expert in both Western and South-East Asian cuisines

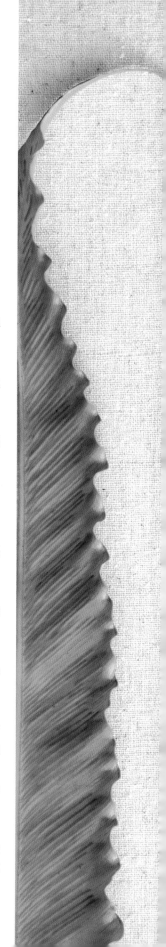

開門9件事
從基本調味料煮起

The Magnificent 9
basic seasonings for wonderful dishes

作者　Author
黃婉瑩　Annie Wong

策劃/編輯　Project Editor
Catherine Tam

攝影　Photographer
Imagine Union

美術統籌及設計　Art Direction & Design
Amelia Loh

出版者　Publisher
Forms Kitchen
香港鰂魚涌英皇道1065號　Room 1305, Eastern Centre, 1065 King's Road,
東達中心1305室　Quarry Bay, Hong Kong
電話　Tel:　2564 7511
傳真　Fax:　2565 5539
電郵　Email: info@wanlibk.com
網址　Web Site:　http://www.formspub.com
　　　　http://www.facebook.com/formspub

瀏覽網站

會員申請

發行者　Distributor
香港聯合書刊物流有限公司　SUP Publishing Logistics (HK) Ltd.
香港新界大埔汀麗路36號　3/F., C&C Building, 36 Ting Lai Road,
中華商務印刷大廈3字樓　Tai Po, N.T., Hong Kong
電話　Tel:　2150 2100
傳真　Fax:　2407 3062
電郵　Email: info@suplogistics.com.hk

承印者　Printer
中華商務彩色印刷有限公司　C & C Offset Printing Co., Ltd.

出版日期　Publishing Date
二○一三年七月第一次印刷　First print in July 2013
二○一七年一月第三次印刷　Third print in January 2017